Green Chemistry for Sustainable Water Purification

Scrivener Publishing
100 Cummings Center, Suite 541J
Beverly, MA 01915-6106

Publishers at Scrivener
Martin Scrivener (martin@scrivenerpublishing.com)
Phillip Carmical (pcarmical@scrivenerpublishing.com)

Green Chemistry for Sustainable Water Purification

Edited by
Shahid-ul-Islam
Abid Hussain Shalla
and
Mohammad Shahadat

Scrivener Publishing

WILEY

This edition first published 2023 by John Wiley & Sons, Inc., 111 River Street, Hoboken, NJ 07030, USA and Scrivener Publishing LLC, 100 Cummings Center, Suite 541J, Beverly, MA 01915, USA

For more information about Scrivener publications please visit www.scrivenerpublishing.com.

Wiley Global Headquarters
111 River Street, Hoboken, NJ 07030, USA

For details of our global editorial offices, customer services, and more information about Wiley products visit us at www.wiley.com.

Library of Congress Cataloging-in-Publication Data

ISBN 978-1-119-85229-2

Cover image: Pixabay.Com
Cover design by Russell Richardson

Set in size of 11pt and Minion Pro by Manila Typesetting Company, Makati, Philippines

Printed in the USA

10 9 8 7 6 5 4 3 2 1

Contents

Preface

Living beings regard water as the most important component in sustaining their lives. It is a vital source of human life because of its organic minerals, essential metal ions, and other caloric sources. The presence of minerals in water is essential for the biological activity of living beings. Water has the potential to transform into various forms, and the discharge of industrial waste into bodies of water without pre-treatment contaminates water, which becomes the main cause of water pollution. On the other hand, drinking pollutant-containing water results in the disturbance of biological activities and a number of disorders and diseases in living beings. People around the globe are facing a crisis of safe portable water and more than three billion people still need appropriate drinking water. It is expected that a huge water crisis will be faced by 2065 owing to the expansion of the population and industrialization worldwide. There, it is necessary to treat industrial waste before releasing it into any body of water. Today, pre-treatment of wastewater has been performed using physical, biological, and chemical treatment methods, which include filtration, anaerobic treatment, solar disinfection, reverse osmosis, oxidation-reduction, plasma treatment, and clay-based low-cost adsorbent nanocomposite materials. Among various treatment techniques, green chemistry, being an environmentally-friendly technology with ease of operation and significant pollutant removal capacity, shows the most significant academic research interest.

This book discusses the different treatment technologies with a special focus on the green adsorption approach, using biological and hybrid biochemical treatment technologies to prevent water contamination and maintain the eco-system. The book discusses the analysis of organic and inorganic pollutants from industrial wastewater. It also focuses on the removal and recovery of organic and inorganic contaminants from the environment. Several case studies describing the removal and recovery of environmental pollutants using green technology are an attractive feature of the present book. The recycling of low-cost along with green adsorbent technology is explained in detail. Finally, the book highlights treatment technologies with effective pollutant removal capacities that are used in modern water treatment units.

We are highly thankful to all the contributing authors for accepting our invitation and contributing their valuable projects in the form of book chapters. All the book chapters are well written, updated with the recently published literature, and we hope the content will be very useful for the researchers, environmental scientists, engineers, and students directly involved with wastewater treatment technology.

Editors
October 2022

1

Green Chemistry for Water Remediation

Syed Wazed Ali*, Satyaranjan Bairagi and Swagata Banerjee

Department of Textile and Fiber Engineering, Indian Institute of Technology Delhi, Hauz Khas, New Delhi, India

Abstract

With the curve of industrialization growing exponentially, the quality index of natural resources is facing a deteriorating trend. Water is one of the most abundantly available natural resources that is being exploited tremendously by humans. Gallons of water are consumed in our day-to-day activities, starting from industrial to household. This exploitation has proved to be a boomerang for them and, hence, ways are being sought to recover, reuse, and preserve this natural resource. Treatment of wastewater has been a subject of interest for a long time, and techniques have been established for the same. However, the chemical treatments of wastewater often adversely affect the environment. The by-products of these processes are often toxic and pose serious health hazards. Hence, a concept like green chemistry has come into the picture for the remediation of water resources. It has proved to be one of the key tools to achieve sustainability by providing appropriate solutions to existing problems. It helps to avoid the toxic by-products of conventional techniques and enhances ecofriendly wastewater treatment approaches. This chapter deals with the various principles of green chemistry in brief and the methods of water remediation in detail. Various chemical treatments of water using green technology have been discussed in detail. Various challenges faced in the path of treatment have also been highlighted.

***Keywords*:** Green chemistry, water remediation, eco-friendly

**Corresponding author*: wazed@iitd.ac.in

Shahid-ul-Islam, Abid Hussain Shalla and Mohammad Shahadat (eds.) Green Chemistry for Sustainable Water Purification, (1–20) © 2023 Scrivener Publishing LLC

1.1 Introduction

It has been stated by the World Health Organization (WHO) that a person consumes about 50 to 100 liters of water daily to fulfil their essential needs and maintain their health. The right to acquire a sufficient and non-toxic quantity of drinking water applies to every living being in our society ignoring what financial status and society they belong to. It has been surveyed that 13.6% of the world's population is not able to access safe drinking water [1] and on the other hand 2.6 billion people face a shortage of the required hygiene. Earlier in the 21st century, around 600 km^3 per year of water was consumed by people staying throughout the world, whereas during the start of this century, the total consumption of drinking water has been elevated to 5300 km^3 per year. It has been surveyed that after the year 2060, there will be some stabilization with four billion human beings staying in city areas proving a very high demand for water consumption [2]. Furthermore, the future prediction states that around 1.8 billion living beings will be facing a shortage of water and approximately 66% of the whole population worldwide will live with the problem of water shortage from the year 2025 [3]. It has also been observed that the main cause for such a disaster is the pollution generated by various industries on our natural water resources. The various wastes from industries, households, agriculture, and farming adversely affect our worldwide drinking water resources.

In the upcoming era, the management of water quality can be done by refining the raw water available in our nature, improving the technology related to such raw water treatment, and properly maintaining the network related to the drinking water supply. Also, it has been stated by many researchers that managing, improving, and inventing new technologies related to water supply can give rise to ways that can achieve efficient and pure quality water supply. In this context, scientists have come up with methodologies that can improve the quality of water by using less toxic or environmentally friendly approaches, some of which are elaborated on in the upcoming section of this chapter. Researchers have also added in their studies that such green approaches give rise to improved and less toxic chemicals, minimization of wastes, and less consumption of energy resources [4]. Therefore, from these approaches, the idea of introducing green chemistry for water purification has approached the minds of various scholars. The idea of such introduction to green chemistry is termed by 12 different principles, which can be stated as follows: preventing the generation of wastes, incrementing the rate of reaction conversions, developing less toxic chemicals, developing safe products made with much less toxic chemicals, increasing efficient energy technologies, utilizing renewable feedstocks, eliminating toxic derivatives of chemicals, utilizing catalysts to increase the rate reactions, designing biodegradable chemical-based products, analyzing ways to diminish pollution, and reducing the causes that lead to accidents [5].

1.2 Challenges in Water Remediation

Several green movements have objected to the use of chemicals in water treatment that harms the environment. Some ask for abolishing the use of such chemicals while others aim at replacing them. Chlorine is one such chemical that is used for disinfecting water while having harmful impacts on the surroundings at the same time [6, 7]. Chlorine kills the microorganisms responsible for diseases like cholera, typhoid, hepatitis, etc. At the same time, however, it is responsible for the formation of compounds like chloroform that are carcinogenic. Disinfecting with chloramines helps to prevent the formation of chloroform, but the cyanogen chloride formed as a by-product is toxic. Several reagents have been tried to eliminate the chloroform formation [8].

Water has an organic matter of natural origin and synthetic origin. The naturally present organic matter in water is called natural organic matter while the organic matter added to water due to human activities is called synthetic organic matter. The percentage of synthetic organic matter is less compared to that of natural organic matter; however, synthetic organic matter is more harmful compared to natural organic matter. The Blackfoot disease is a well-known disease in Taiwan that causes numbness and coldness of limbs [9, 10]. It is thought to be due to the presence of fluorinated compounds in water. Humic acid is one such fluorinated compound that causes organ-related diseases. It is a high molecular weight complex molecule that negatively impacts the erythrocytes via the generation of reactive oxygen species [11]. Metals like copper and aluminum are also related to toxicity issues [12, 13].

1.3 Green Chemistry as a Novel Alternative to Conventional Wastewater Treatment

1.3.1 Green Chemistry

The conceptualization of green chemistry was a result of the Pollution Prevention Act, in the year 1990 [4]. It stated the US national policy to eradicate pollution by improvising its design and processes rather than the treatment and disposal of products. In other words, the term green chemistry refers to the design and development of products that would help to reduce or get rid of the production of any hazardous substances [14]. There exists a small line of difference between environmental chemistry and green chemistry. Environmental chemistry deals with environmental pollutants. Green chemistry, on the other hand, aims at designing new techniques and improving existing technologies that would minimize the generation of waste. Green chemistry also includes certain crucial ideas. One is the impact of all raw materials used in product synthesis and not just the principal raw material. The other is the optimization of an efficient process with the least negative impacts. Green chemistry is based on twelve principles that may be stated as under. The "Twelve Principles" related to "Green Chemistry" were introduced in the year 1998 [5]. These principles provide a strong guideline to manufacture ways of different less harmful chemical products and technologies, which apply to almost all life cycle processes, including the raw materials utilized for enhancing the productivity and safeness of the principles used, along with the toxic nature and degradability of different products and chemicals used in this green chemistry approach. Some of them are listed as prevention, atom economy, less hazardous chemical synthesis, designing safer chemicals, safer solvents, and auxiliaries, design for energy efficiency, use of renewable feedstocks, reducing derivatives, catalysis, design for degradation, real-time analysis for pollution prevention and inherently safer chemistry for accident prevention.

Proper waste handling: It aims to prevent waste generation rather than treat and clean them up when produced. It is one of the most important of the twelve different principles of this green chemistry approach. It is always safer to avoid the accumulation of waste other than to clean up after it has been formed. Waste can be defined as anything that has no appreciated importance or the product produced from any damage caused to any energy technology. Also, it has been stated in many studies that wastes can take different forms and can cause an impact on nature in a different way depending on the waste's nature, toxicity, quantity, or the way it has been drained out [15]. When huge quantities of the precursors are utilized in a method of synthesis, then much of them are lost due to the original design of the procedure itself and this generation will give rise to undesirable wastes. One of such examples to minimize waste quantity is to use molecular oxygen, thereby eliminating the requirement for chlorine. This has dropped the generation of wastes close to 0.3 kg of than when done by the conventional method. Scientists have reported that this new method can produce more than 16 times less amount of waste than the conventional method, thereby removing the generation of wastewater [16]. It has also been reported in the literature that when we are not able to avoid by-products, then another new approach must be taken into consideration and an approach that is productive to search for an environment-friendly industrial process that can cause waste to transform into a newer type of material with an important value for other processes as it can reenter our life cycle. This method is used for the generation of biofuel (Figure 1.1).

Atom economy: It aims to include all raw materials in the synthesized product. In the year 1990 Barry Trost established the idea of using the efficiency of synthetics i.e. Atom Economy abbreviated as A.E., also known as Atom Efficiency [17]. The concept refers to maximizing the utilization of precursors which can cause the final end-products to have more atoms prevailing from the reactant side. The perfect reaction will have all the atoms present on the reactant side. The atomic efficiency is estimated as the ratio of molecular weights of the product to that of all the reactants

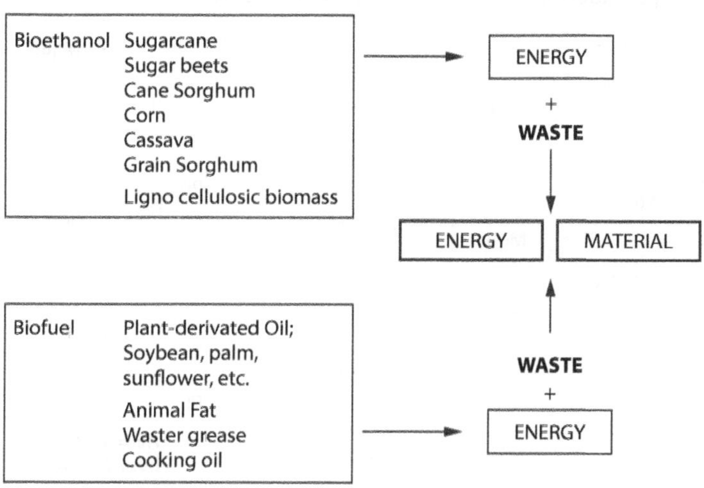

Figure 1.1 Generation of biofuel by-products [5].

used in this reaction, as shown in Figure 1.2. AE is a theoretical value used to evaluate the process by which a reaction will take place efficiently. A few examples of such reactions are Grignard, A3 coupling, and the Diels–Alder reaction. Figure 1.2 shows a typical Grignard reaction and application of the Grignard reagent to create a propargylic amine type of structure. The AE value for such a reaction is 56%, which indicates losing half amount of the precursor used in this reaction.

Safe reaction: The reactions should be so designed that they have little or no toxic effect on the environment. As shown in Figure 1.3, many new works in this field of green chemistry have improved the present scenario of causing water remediation. The reactions involved in such a green chemistry approach are modifications of reactions that have been invented a long time back. Less hazardous reactions involving cycloadditions [17] rearranging or multiple components using coupling type of reactions [18] were known long back and comprise one efficient category to provide green chemical reactions. Cascade type or tandem type of reactions [19], carbon and hydrogen activation [20] metathesis [21] and reactions involving enzymes [22, 23] are innovative ways to depict efficient examples of more clean and less toxic synthetic tools present to different organic chemists.

Safe chemicals: The chemicals should be designed so that they carry out their functionalities without any toxic side effects. This principle for providing efficient green chemistry is to design chemical products that do not

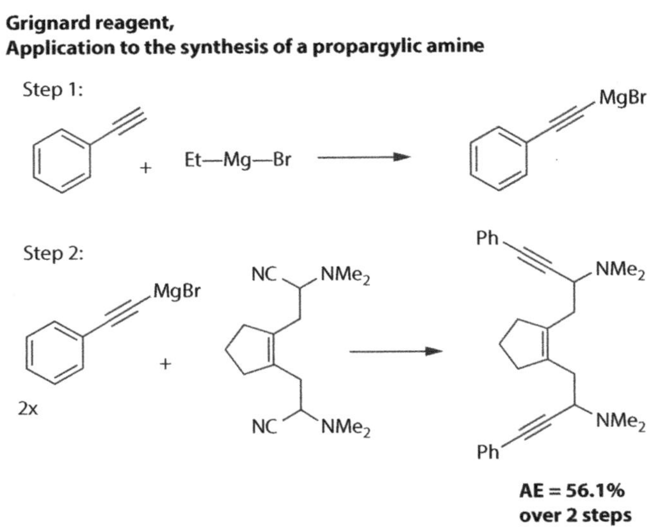

Figure 1.2 Atom economy for Grignard reagent [5].

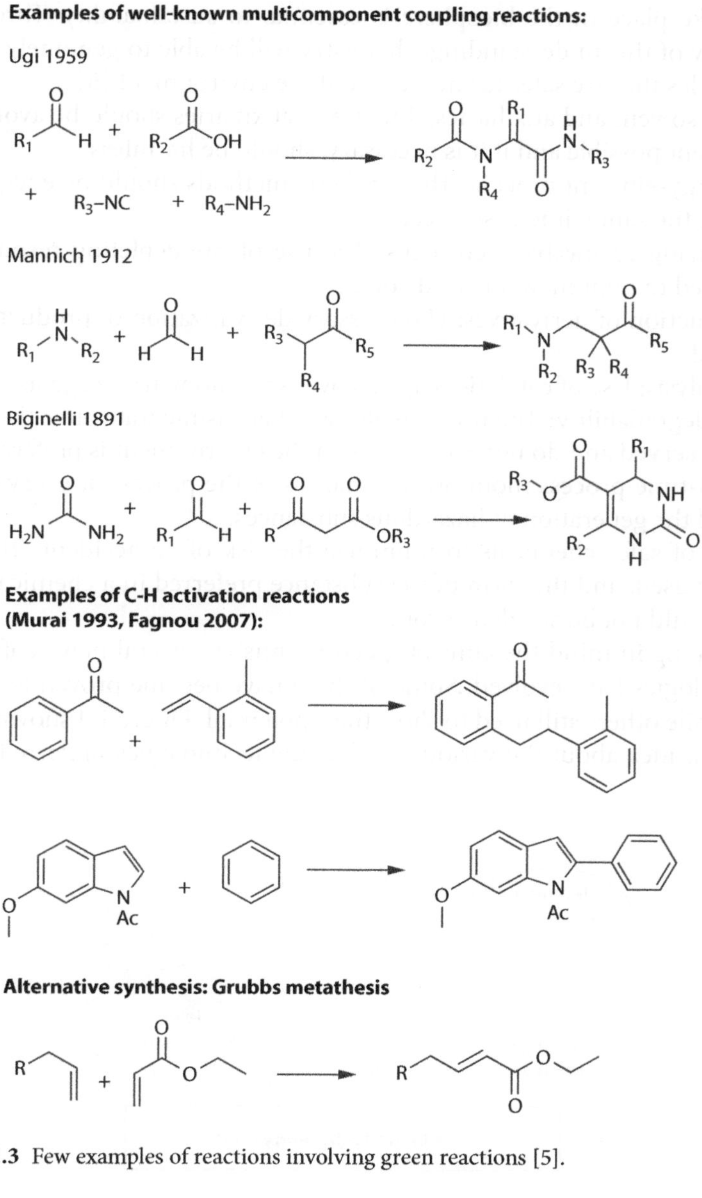

Figure 1.3 Few examples of reactions involving green reactions [5].

exhibit toxicity to our environment. Such designed structures offer different functions in the field of medicines to materials, which are less toxic than the conventional made materials. Understanding the properties of a molecule that have an impact on the environment and the transformations that take place in the biosphere is essential to sustainability. Through a mastery of this understanding, chemistry will be able to genuinely design molecules that are safer for humans and the environment [5].

Safe solvent and auxiliaries: The use of auxiliaries should be avoided to the extent possible and if it is necessary, should be harmless.

Energy-efficient process: The synthetic methods should be energy efficient at the same time cost-effective.

Utilizing renewable feedstocks: The use of nondepleting resources is preferred to nonrenewable feedstocks.

Reduction of derivatives: Unnecessary derivatization of products to be avoided.

Catalysis: Use of catalytic reagents over stoichiometric reagents.

Biodegradability: The use of materials that disintegrate after their purpose is served and do not accumulate in the environment is preferred.

Real-time process monitoring: Analysis of the process at every step to control the generation of hazardous substances.

Use of safer chemicals to minimize the risk of an accident: The substances used, and the form of the substance preferred in a chemical pathway should not be accident-prone.

Keeping in mind the aims of green chemistry, several new ecofriendly technologies have evolved. Some of them have become proven technologies while others still need to show their potential. Figure 1.4 shown below gives an idea about the various ecofriendly technologies preferred in use

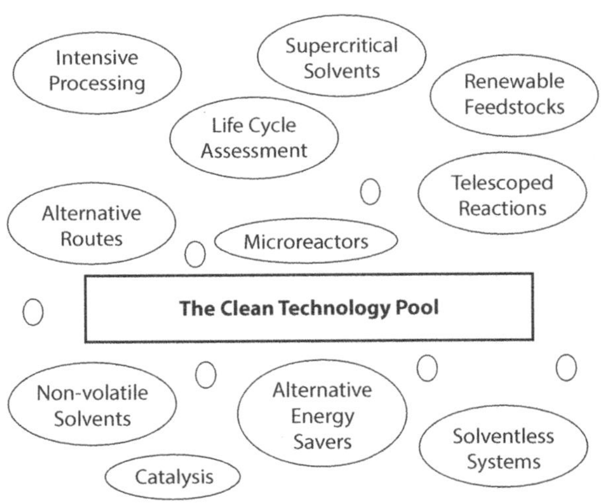

Figure 1.4 The ecofriendly technologies in use [4].

today. Advancements concerning the synthesis of materials have also been carried out. Special reactors like continuous flow reactors, microchannel reactors have been designed to make processes more efficient and eco-friendly. These reactors may be combined with the green chemistry methods to design improved technologies that aim at providing better results.

1.3.2 Applications of Green Chemistry in Water Remediation

Conventional water treatment technologies are associated with the generation of highly toxic chemicals that affect the land, water, and air ecosystems. Sustainability being the need of the hour, novel green solutions are being preferred and searched for the treatment of this nonrenewable valuable natural resource. These green technologies seek to curb the overexploitation and misuse of water resources in the first place. The green remediation methods eliminate the chances for the liberation of highly toxic by-products that would otherwise harm the environment [24]. In this manner, they meet the needs of the present without compromising the needs of future generations. Some of the green technical approaches for water remediation are discussed below:

Advanced oxidation processes: This is a chemical treatment of water that proceeds with oxidation reactions. This process seeks to remove organic compounds from wastewater with the help of some highly reactive species. They have also been used for the removal of dyes from effluents [25, 26]. The advanced oxidation processes may be grouped into two based on the use of UV light for the procedure, namely nonphotochemical and photochemical advanced oxidation processes.

The nonphotochemical methods produce reactive hydroxyl radicals in the absence of light through processes like ozonation, Fenton's reagent oxidation, wet air oxidation, and electrochemical oxidation. The ozonation process is a widely preferred ecofriendly process that helps to break down the chromophoric groups in organic compounds into smaller molecules. This process is carried out in three steps consisting of ozone generation, dissolution of ozone in the wastewater, and finally oxidation of organic compounds in wastewater [27]. Ozone is a strong oxidizing agent. It indulges in slow chemical reactions which are driven kinetically than by thermodynamics. As far as water treatment is concerned, direct ozonation is a selective and slow process wherein the compounds in water containing certain functional groups are only attached by ozone. For instance, ozone is not preferred for the removal of compounds like hydrocarbons (alkanes) or chlorinated organic compounds. They are used for the oxidation of phenolic compounds and polyaromatic hydrocarbons [28]. In such situations, ozone is used along with some homogenous and heterogeneous catalysts that help to promote the degradation of the organic compounds [29]. The removal of dyes from aqueous media through ozonation has also been established as a viable technique [30, 31]. Another oxidation-based process is Fenton's reagent oxidation (Figure 1.5). The Fenton's reagent comprises a mixture of chemical oxidizers, hydrogen peroxide, ferrous ions as the catalyst, and an acid source for maintaining pH. The easy availability and nontoxicity of the constituents make this reagent a cost-effective oxidizing agent. The ferrous ion catalyst act on hydrogen peroxide to produce hydroxyl radicals through several cyclic reactions [32]. The reaction mechanism is explained as below [33]:

$$Fe^{2+} + H_2O_2 \longrightarrow Fe^{3+} + HO^{\cdot} + OH^{-}$$

$$HO^{\cdot} + H_2O_2 \longrightarrow HO_2^{\cdot} + H_2O$$

$$Fe^{2+} + HO^{\cdot} \longrightarrow Fe^{3+} + OH^{-}$$

$$Fe^{3} + HO_2^{\cdot}\ Fe^{2+} + O_2 + H^{+}$$

$$HO^{\cdot} + HO^{\cdot} \longrightarrow H_2O_2$$

$$Organic\ pollutant + HO^{\cdot} \longrightarrow Degraded\ products$$

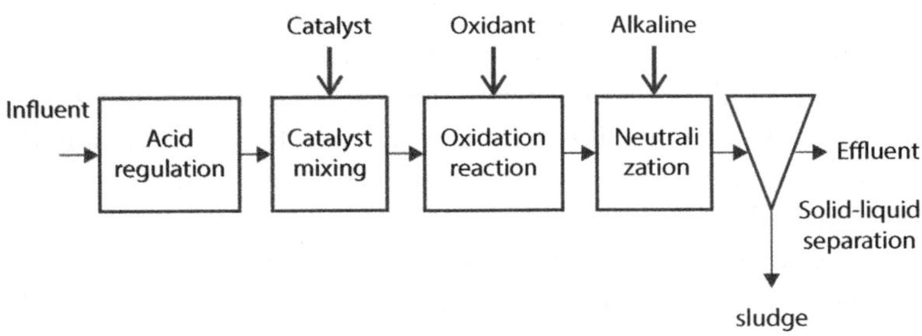

Figure 1.5 Illustration of wastewater treatment by Fenton's oxidation process [32].

The presence of UV light further helps to increase the process efficiency. This is then called the photo-Fenton process. The equations have been shown below [33]:

$$Fe^{3+} + H_2O + hv \longrightarrow Fe^{2+} + HO^{\cdot} + H^{+}$$

$$Fe^{3+} + H_2O + hv \longrightarrow Fe^{2+} + HO^{\cdot} + H^{+}$$

The Fe^{3+} ions complexes generate additional hydroxyl radicals and regenerate Fe^{2+} ions for further reaction with hydrogen peroxide. This reagent has also been used for the removal of dye from effluents [34, 35].

The wet air oxidation process also removes the organic contaminants from wastewater with the help of oxygen or air at high temperatures and pressure. In this process, the contaminants are broken down into biodegradable substances with the release of harmless gases like carbon dioxide and nontoxic by-products like water and inorganic salts. This treatment comprises a physical phase and a chemical phase. The diffusion of oxygen from gas to the liquid and the release of carbon dioxide from the liquid into the gaseous phase is part of the physical phase. The reactions concerning the degradation of organic matter refer to the chemical phase [36, 37]. The electrochemical model of oxidation is another advanced oxidation process used for the treatment of organic pollutants in water. It is electrogenerated *in situ* hydroxyl radicals that help to disintegrate the organic matter present in the wastewater. The electrodes of the reaction serve as the sites for the generation and consumption of electrons. The shape and material of the construction of the electrodes exert a great influence on the electrochemical process of reduction [38]. Electrocoagulation is another green method that is employed for the treatment of wastewater. It mainly deals with the generation of coagulants in situ. These coagulants may be generated from the dissolution of aluminum or iron ions from their respective electrodes. The metal ions are released at the anode while hydrogen gas is liberated at the cathode. This hydrogen gas helps to keep the flocculants afloat in water. A network is formed by the liberated metal ions that help in the chemical adsorption of the contaminants. This whole process is referred to as electrocoagulation [39]. A certain amount of metal ions are required for the treatment of a given level of polluted wastewater. Iron being cheaper than aluminum is preferred for the treatment of wastewater and aluminum for treating water. Textile wastewater [40] and municipal sewage [41] have also been treated through electrocoagulation. Electrofloatation is another process that involves the water treatment process by helping to keep the pollutants afloat. The release of oxygen and hydrogen gases at the electrodes through water electrolysis forms tiny bubbles that float the pollutants at the water surface. Hydrogen is released at the cathode and oxygen is liberated at the anode. The size of the bubbles formed also affects the removal efficiency of the pollutants. The bubble size in turn depends on the pH of the solution [42]. While hydrogen bubbles are small at neutral pH, the size of the oxygen bubbles increases as the pH increases. The smaller the size of the bubbles, the higher is the surface area provided. This helps to increase the separation efficiency of the electro flotation method. Hence the efficacy of the electrolocation process increases with the tiny bubbles with greater uniformity. Other factors like electrode composition, cell design, and water conductivity also affect the electrolocation process [43].

Another category of advanced oxidation processes is the photochemical advanced oxidation process. This category of processes deals with the principles of photochemistry. The reactions thus are carried out at much lower temperatures and with processes being substratesspecific [44]. Photocatalytic oxidation is one such method of wastewater treatment. Photocatalysts like TiO_2 (Figure 1.2), ZnO, Fe_2O_3, and CdS, mainly semiconductors are used to disintegrate organic contaminants in the presence of light energy. TiO_2, however, is usually preferred for photocatalytic degradation due to its availability, low cost, nontoxicity, and appreciable physical and chemical properties [45]. To enhance the photocatalytic activity of these photocatalysts, they are sometimes immobilized onto substrates like activated carbon, zeolites, which provide a large surface area for adsorption. It provides further advantages of leaching prevention and photocatalyst recovery [46].

Another advanced oxidation process becoming popular nowadays is the oxidation by hydrogen peroxide in the presence of UV light. Hydrogen peroxide generates hydroxyl radicals in the presence of UV light. Hydrogen peroxide alone can remove pollutants in water. However, it is unable to remove complex contaminants by the process of oxidation (Figure 1.6). The activity and efficiency of hydrogen peroxide are enhanced in the presence of a light energy source that enables it to generate more hydroxyl radicals. This method has also been used for the treatment of textile dye effluents [48, 49]. With a similar approach, ozonation has been combined with UV light sources to enhance the effectiveness in the removal of organic pollutants in wastewater. UV assists in improving ozone decomposition that increases the yield of hydroxyl radicals, in turn enhancing the ozonation rate. The initiation of the reaction in the case of photocatalytic ozonation is

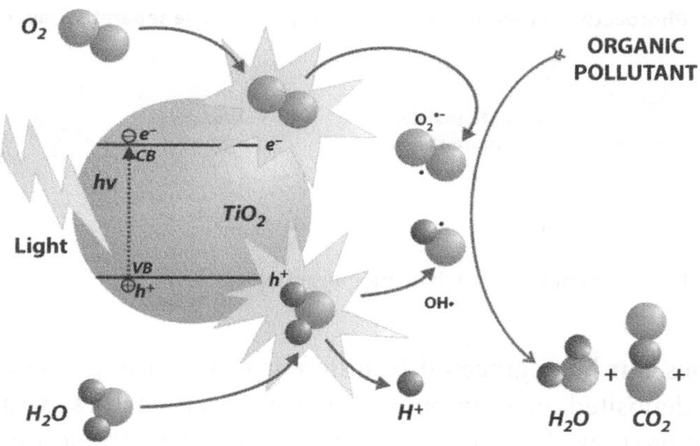

Figure 1.6 Mechanism of TiO_2 photocatalysis [47].

by the electron transfer from photocatalyst to oxygen forming starting radical. This mechanism differs for ozonation alone where the starting radical is formed by the reaction of hydroxyl ions and ozone [50].

A novel technique to decontaminate wastewater is through the absorption of pollutants on the surface of a catalyst [51]. Catalysts like activated carbon, silica gel, etc. are used as the substrate for the adsorption of contaminants. Gradually, the implementation of nanoparticles emerged in this field. The advantage of high surface area was also realized for the adsorption of the contaminants. Magnetic nanoparticles bearing the structural formula AB2O4 were used for this purpose [52, 53]. They provided an additional advantage of the catalyst recovery through magnetic separation, which is otherwise difficult with the other adsorbents. Magnetic nanoparticles are used for preferential adsorption of arsenic ions followed by magnetic decantation [54].

There are also membrane processes that are used for wastewater treatment [55]. The membrane processes offer advantages of low chemical and energy consumption, good stable water quality, which does not depend on the quality of the water to be treated, easy to maintain technology with a good scalability potential. The membranes may differ in their characteristics, ranging from organic to inorganic, porous or dense, electrically charged or neutral, etc. Based on this, the driving force of the mechanism of treatment may be pressure-driven or concentration gradient of electric potential difference. The photocatalytic membrane reactors are a membrane technology process where photocatalysis is combined with membrane technology, the driving force being the pressure difference. These

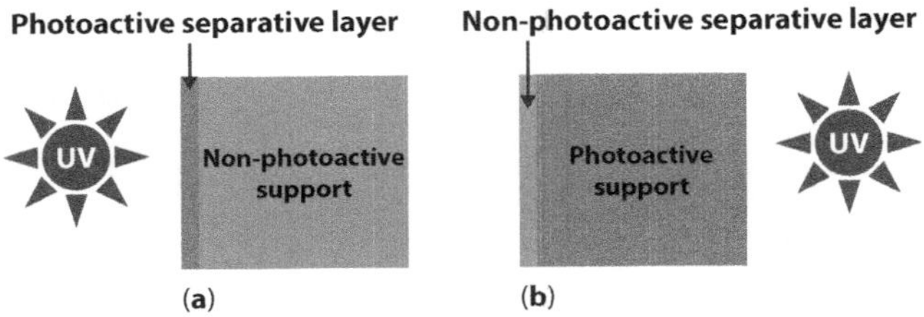

Figure 1.7 Two different photocatalytic membranes [56].

membranes can be engineered in different ways. In one case a photoactive layer deposited on a nonphotoactive layer acts as the separating layer. The nonphotoactive layer acts as the supportive layer. The second way is to deposit a nonphotoactive layer on a photoactive layer where the nonphotoactive layer serves as the separating layer (Figure 1.7).

Membrane bioreactor technology couples a biological process with a membrane separation step [57]. This is another green method for wastewater treatment. It offers advantages like a decrease in sludge production, reduced footprint, etc. compared to the conventional treatment procedures. The cost of membrane fabrication and the energy demand of this process leads to its reconsideration as a green technology [24]. Modifications of the method have been carried out to render them energy efficient [58, 59].

1.4 Conclusion

Water remediation has become the need of the hour if we are to save our future generations from facing a water crisis. The existing water treatment techniques aim to purify water out of its contaminants, releasing some harmful by-products at the same time. These by-products in turn affect the environment and its surviving species. Thus, water technologists and chemists have evolved with green techniques for treating water. These are based on the fundamental principles of green chemistry. A total of 12 principles are included under green chemistry that has been discussed in detail in this chapter. Novel green water remediation techniques have been dealt with in detail in this chapter. Starting from advanced oxidation processes to adsorption techniques, extending to membrane technologies have been explored for their water remediation potentials. These are some of the novel green methods of water decontamination that serve the purpose to the fullest without any adverse effect on the surroundings. These methods are therefore gaining attention nowadays for the treatment of water.

Modifications of these methods are also underway, which will serve as green technologies for water remediation shortly.

Acknowledgment

The authors are thankful to the Indian Institute of Technology Delhi (IITD), India, forgiving platform to write this chapter.

References

1. WHO, The Human Right to Water and Sanitation Media brief, UN-Water Decad. Program. *Advocacy Commun. Water Supply Sanit. Collab. Counc.*, 1–8, 2011. http://www.un.org/waterforlifedecade/pdf/human_right_to_water_and_sanitation_media_brief.pdf.
2. Jain, S. and Singh, V., *Water resources systems planning and management*, Elsevier, 51, 3–858, 2003, https://books.google.co.in/books?hl=en&lr=&id=dR1sFWlUpPUC&oi=fnd&pg=PP1&dq=Jain+and+Singh,+2003,+Water+Resources+Systems+Planning+and+Management,+Volume+51,+3-858&ots=n-l81Tpvel3&sig=l9khk3m1ZH2K2AayDmnhGZOTNxg.
3. Pritchard, A., Statistical bibliography or bibliometrics. *J. Doc.*, 25, 348, 1969.
4. Ghernaout, D., Ghernaout, B., Naceur, M.W., Embodying the chemical water treatment in the green chemistry-A review. *Desalination*, 271, 1–10, 2011, https://doi.org/10.1016/j.desal.2011.01.032.
5. Anastas, P. and Eghbali, N., Green chemistry: Principles and practice. *Chem. Soc Rev.*, 39, 301–312, 2010, https://doi.org/10.1039/b918763b.
6. Gribble, G.W., *Natural production of organohalogen compounds*, Springer, 2003, http://www.springerlink.com/index/10.1007/b10466.
7. Gordon, G.W., of Natural production chlorinated compounds. *Environ. Sci. Technol.*, 28, 310–319, 1994.
8. Stefan, M.I., *Advanced oxidation processes for water treatment*, I. Mihaela (Ed.), IWA Publishing, 2019.
9. Cheng, M.L., Ho, H.Y., Chiu, D.T.Y., Lu, F.J., Humic acid-mediated oxidative damages to human erythrocytes: A possible mechanism leading to anemia in Blackfoot disease. *Free Radic. Biol. Med.*, 27, 470–477, 1999, https://doi.org/10.1016/S0891-5849(99)00072-6.
10. Tseng, C.H., Blackfoot disease and arsenic: A never-ending story. *J. Environ. Sci. Heal. - Part C Environ. Carcinog. Ecotoxicol. Rev.*, 23, 55–74, 2005, https://doi.org/10.1081/GNC-200051860.
11. Gau, R.J., Yang, H.L., Suen, J.L., Lu, F.J., Induction of oxidative stress by humic acid through increasing intracellular iron: A possible mechanism leading to

atherothrombotic vascular disorder in Blackfoot disease. *Biochem. Biophys. Res. Commun.*, 283, 743–749, 2001, https://doi.org/10.1006/bbrc.2001.4832.
12. Beccaria, A., Lahiri, D.K., Bondy, S.C., Chen, D.M., Hamadeh, A., Li, H., Taylor, R., Campbell, A., Aluminum and copper in drinking water enhance inflammatory or oxidative events specifically in the brain. *J. Neuroimmunol.*, 176, 16–23, 2006, https://doi.org/10.1016/j.jneuroim.2006.03.025.
13. Levesque, L., Ligand specific effects on aluminum incorporation and toxicity. pdf. *Brain Res.*, 877, 191–202, 2000.
14. Sinha, P., Sustainable textiles. *Radar*, 2, 19–21, 2011, https://doi.org/10.5920/radar.2011.2119.
15. Sheldon, R.A., E factors, green chemistry and catalysis: An odyssey. *Chem. Commun.*, 29, 3352–3365, 2008, https://doi.org/10.1039/b803584a.
16. Kilty, P.A. and Sachtler, W.M.H., Catalysis reviews: Science and engineering, in: *The mechanism of the selective oxidation ethylene to ethylene*, vol. n.d., pp. 37–41.
17. Trost, B.M., On inventing reactions for atom economy. *Acc. Chem. Res.*, 35, 695–705, 2002, https://doi.org/10.1021/ar010068z.
18. Yavari, I., Nematpour, M., Hossaini, Z., A synthesis of phosphorylated dioxohexahydropyrimidines from N,N′-dimethylurea, activated acetylenes, and trialkyl phosphites. *Mol. Divers.*, 14, 617–620, 2010, https://doi.org/10.1007/s11030-009-9193-1.
19. Nicolaou, K.C., Montagnon, T., Snyder, S.A., Tandem reactions, cascade sequences, and biomimetic strategies in total synthesis. *Chem. Commun.*, 3, 551–564, 2003, https://doi.org/10.1039/b209440c.
20. Labinger, J.A. and Bercaw, J.E., Understanding and exploiting C-H bond activation. *Nature*, 417, 507–514, 2002, https://doi.org/10.1038/417507a.
21. Grubbs, R.H., Olefin metathesis. *Tetrahedron.*, 60, 7117–7140, 2004, https://doi.org/10.1016/j.tet.2004.05.124.
22. Dannenberg, J.J., Book reviews, *J. Am. Chem. Soc.*, 123, 1009–1011, 1998.
23. Cavell, K., Golunski, S., Miller, D., Handbook of green chemistry - green catalysis. *Platin. Met. Rev.*, 54, 233–238, 2010, https://doi.org/10.1595/147106710x527928.
24. Azimi, S.C. and Pendashteh, A., Green technologies for wastewater treatment. *Green Technologies for Wastewater Treatment Second International Conference in New Research on Chemistry & Chemical Engineering*, 2018.
25. Atalay, S. and Ersöz, G., Advanced oxidation processes for removal of dyes from aqueous media. *Green Chem. Dye. Remove. Wastewater Res. Trends Appl.*, 83–117, 2015, https://doi.org/10.1002/9781118721001.ch3.
26. Nidheesh, P.V., Zhou, M., Oturan, M.A., An overview on the removal of synthetic dyes from water by electrochemical advanced oxidation processes. *Chemosphere*, 197, 210–227, 2018, https://doi.org/10.1016/j.chemosphere.2017.12.195.

27. Palit, S. and Studies, E., An overview of ozonation associated with nano-filtration as an effective procedure in treating dye effluents from textile industries with the help of a bubble column reactor. *Int. J. Chem. Sci.*, 10, 27–35, 2016.
28. Peleg, M., The chemistry of ozone in the treatment of water. *Water Res.*, 10, 361–365, 1976, https://doi.org/10.1016/0043-1354(76)90052-X.
29. Kasprzyk-hordern, B., Ziółek, M., Nawrocki, J., Catalytic ozonation and methods of enhancing molecular ozone reactions in water treatment. *Appl. Catal. B*, 46, 639–669, 2003, https://doi.org/10.1016/S0926-3373(03)00326-6.
30. Recycling, E., Hu, E., Shang, S., Removal of Reactive Dyes in Textile effluents by Catalytic Ozonation Pursuing on-Site. *Molecules*, 19, 1–21, 2019.
31. Sevimli, M.F. and Sarikaya, H.Z., Ozone treatment of textile effluents and dyes: Effect of applied ozone dose, pH and dye concentration. *J. Chem. Technol. Biotechnol.*, 842–850, 2002, https://doi.org/10.1002/jctb.644.
32. Xu, M., Wu, C., Zhou, Y., Advancements in the Fenton process for wastewater treatment. *Adv. Oxid. Process. - Appl. Trends, Prospect.*, 1–17, 2020, https://doi.org/10.5772/intechopen.90256.
33. Ameta, R., Chohadia, A.K., Jain, A., Punjabi, P.B., Fenton and photo-Fenton processes, pp. 49–87, Elsevier, 2018, https://doi.org/10.1016/B978-0-12-810499-6.00003-6.
34. Patil, A.D. and Raut, P.D.P.D., Treatment of textile wastewater by Fenton's process as a advanced oxidation process. *IOSR J. Environ. Sci. Toxicol. Food Technol.*, 8, 29–32, 2014, https://doi.org/10.9790/2402-081032932.
35. Nidheesh, P.V., Gandhimathi, R., Ramesh, S.T., Degradation of dyes from aqueous solution by Fenton processes: A review. *Environ. Sci. Pollut. Res.*, 20, 2099–2132, 2013, https://doi.org/10.1007/s11356-012-1385-z.
36. Levec, J. and Pintar, A., Catalytic wet-air oxidation processes: A review. *Catal. Today*, 124, 172–184, 2007, https://doi.org/10.1016/j.cattod.2007.03.035.
37. Zhou, M., Särkkä, H., Sillanpää, M., A comparative experimental study on methyl orange degradation by electrochemical oxidation on BDD and MMO electrodes. *Sep. Purif. Technol.*, 78, 290–297, 2011, https://doi.org/10.1016/j.seppur.2011.02.013.
38. Al Khatib, M., Bellini, M., Pogni, R., Giaccherini, A., Innocenti, M., Vizza, F., Lavacchi, A., Effect of electrode shape and flow conditions on the electrochemical detection with band microelectrodes. *Sensors*, 18, 31–39, 2018, https://doi.org/10.3390/s18103196.
39. Chen, G. and Hung, Y., Electrochemical wastewater treatment processes. *Adv. Physicochem. Treat. Technol.*, 5, 57–106, 1946.
40. Ln, S. and Peng, C., Treatment of textile wastewater by electrochemical method. *Water Resour.*, 28, 277–282, 1994.
41. Pouet, M. and Grasmick, A., Electrocoagolation and flotation. *Water Sci. Technol.*, 31, 275–283, 1989, https://doi.org/10.1016/0273-1223(95)00230-K.

42. Alam, R. and Shang, J.Q., Bubble size distribution in a laboratory-scale electroflotation study, *Environ. Monit. Assess.*, 189, 193, 1–14, 2017, https://doi.org/10.1007/s10661-017-5888-4.
43. Reds, T.F., Columns, B., Beds, L.F., Angelino, H., Burckhart, B., Zoll, G., Transfer, M., Stepanek, J.B., Beds, T.F., Worthington, H., Beds, F., Beds, G.F., Nicolais, L., Landel, F., Liquids, N., Transfer, M., Suchozebrski, W., Reds, L.F., Schumpe, A., Yu, Y.H., Lee, S., Kitakyushu, T., Removal of colloidal particles in electroflotation. *AIChE J.*, 31, 201–208, 1985.
44. Bendicho, C., Pena, F., Costas, M., Gil, S., Lavilla, I., Photochemistry-based sample treatments as greener approaches for trace-element analysis and speciation. *Trends Anal. Chem.*, 29, 681–691, 2010, https://doi.org/10.1016/j.trac.2010.05.003.
45. Wang, Y., Sun, C., Zhao, X., Cui, B., Zeng, Z., Wang, A., Liu, G., Cui, H., The application of nano-TiO2 photo semiconductors in agriculture. *Nanoscale Res. Lett.*, 11, 1–7, 2016, https://doi.org/10.1186/s11671-016-1721-1.
46. Velasco, L.F., Haro, M., Parmentier, J., Gadiou, R., Vix-guterl, C., Ania, C.O., Tuning the photocatalytic activity and optical properties of mesoporous TiO ??, in: *Spheres by a carbon scaffold*, 2013.
47. Ibhadon, A.O. and Fitzpatrick, P., Heterogeneous photocatalysis: Recent advances and applications. *Catalysts*, 3, 189–218, 2013, https://doi.org/10.3390/catal3010189.
48. Reagents, U.V.H.O.F., Treatment of textile dyeing wastewater by photo oxidation using treatment of textile dyeing wastewater by photo oxidation using UV / H2O2 / Fe2 + Reagents. *Sci. Asia,*, 32, 181–186, 2016, https://doi.org/10.2306/scienceasia1513-1874.2006.32.181.
49. Al-kdasi, A. and Idris, A., Treatment of textile wastewater by advanced axidation processes – A review. *Global NEST: The Int. J.*, 6, 222–230, 2005.
50. Agustina, T.E., Ang, H.M., Vareek, V.K., A review of the synergistic effect of photocatalysis and ozonation on wastewater treatment. *J. Photoch. Photobio. C*, 6, 264–273, 2006, https://doi.org/10.1016/j.jphotochemrev.2005.12.003.
51. Segneanu, A.E., Orbeci, C., Lazau, C., Sfirloaga, P., Vlazan, P., Bandas, C., Grozescu, I., Wastewater tretment methods. *Water Treat*, 53–80, 2013, http://dx.doi.org/10.5772/53755.
52. Shen, Y.F., Tang, J., Nie, Z.H., Wang, Y.D., Ren, Y., Zuo, L., Preparation and application of magnetic Fe3O4 nanoparticles for wastewater purification. *Sep. Purif. Technol.*, 68, 312–319, 2009, https://doi.org/10.1016/j.seppur.2009.05.020.
53. Gao, F., An overview of surface-functionalized magnetic nanoparticles: Preparation and application for wastewater treatment. *Chem. Select*, 4, 6805–6811, 2019, https://doi.org/10.1002/slct.201900701.
54. Zeng, H., Zhai, L., Qiao, T., Yu, Y., Zhang, J., Li, D., Efficient removal of As(V) from aqueous media by magnetic nanoparticles prepared with Iron-containing water treatment residuals. *Sci. Rep.*, 10, 1–12, 2020, https://doi.org/10.1038/s41598-020-65840-1.

55. Molinari, R., Mungari, M., Drioli, E., Di Paola, A., Loddo, V., Palmisano, L., Schiavello, M., Study on a photocatalytic membrane reactor for water purification. *Catal. Today*, 55, 71–78, 2000, https://doi.org/10.1016/S0920-5861(99)00227-8.
56. Bosc, F., Ayral, A., Guizard, C., Mesoporous anatase coatings for coupling membrane separation and photocatalyzed reactions. *J. Memb. Sci.*, 265, 13–19, 2005, https://doi.org/10.1016/j.memsci.2005.04.039.
57. Artiga, P., Oyanedel, V., Garrido, J.M., Méndez, R., An innovative biofilm-suspended biomass hybrid membrane bioreactor for wastewater treatment. *Desalination*, 179, 171–179, 2005, https://doi.org/10.1016/j.desal.2004.11.065.
58. Kim, J., Kim, K., Ye, H., Lee, E., Shin, C., McCarty, P.L., Bae, J., Anaerobic fluidized bed membrane bioreactor for wastewater treatment. *Environ. Sci. Technol.*, 45, 576–581, 2011, https://doi.org/10.1021/es1027103.
59. Yang, S., Yang, F., Fu, Z., Wang, T., Lei, R., Simultaneous nitrogen and phosphorus removal by a novel sequencing batch moving bed membrane bioreactor for wastewater treatment. *J. Hazard. Mater.*, 175, 551–557, 2010, https://doi.org/10.1016/j.jhazmat.2009.10.040.

2

Advances in Wastewater Treatment Using Natural and Modified Zeolites

Sheikh A. Majid*, **Gowher Jan and Aabid H. Shalla**

Department of Chemistry, Islamic University of Science & Technology, Awantipora, J&K, India

Abstract
Because of their bioaccumulation and poisonous nature, heavy metal ions in wastewater pose a major health risk. There is a variety of traditional methods for removing harmful metal ions from wastewater, but ion exchange/sorption is one of the most successful, cost-effective, and simple to utilize. As sorbents, various materials, such as cellulose, coal, and peat, have been employed in recent years. However, emphasis has been placed on the utilization of low-cost materials as promising sorbents for eliminating hazardous metal ions from wastewater. Aluminosilicate minerals with strong thermal and chemical resistance are known as zeolites. The sorption of various metal ions on different zeolites with variable Si/Al ratios, pore diameters, and surface areas will be the main focus of this chapter. Also discussed are the effects of pH, concentration, temperature, and contact time on adsorption.

Keywords: Zeolites, metal ions, sorption, wastewater, ion exchange

2.1 Global Impact of Wastewater Treatment

The used water discharged from industries, homes, businesses, agriculture, and cities are termed "wastewater" [1]. These countries produce billions of tonnes of wastewater comprising sludge and other waste elements, which can be collected, processed, and ultimately used, either directly or indirectly, or with no benefit [2]. At a global level, evaluations of these

**Corresponding author*: ali.majid66@gmail.com

Shahid-ul-Islam, Abid Hussain Shalla and Mohammad Shahadat (eds.) Green Chemistry for Sustainable Water Purification, (21–48) © 2023 Scrivener Publishing LLC

wastewater streams are limited and sporadic, and there is a general lack of information about them. However, certain international organizations, such as UN-Habitat (2008), FAO/IWMI through AQUASTAT, and Global Water Intelligence (GWI 2014), have made significant efforts to update and renew the evaluations.

Various businesses dump significant amounts of contaminants, both organic and inorganic, into the water. These pollutants, primarily heavy metals, have the potential to be toxic and carcinogenic, causing serious diseases in humans and many other species [3–5]. Metals of high concern include zinc (Zn), lead (Pb), arsenic (As), copper (Cu), cadmium (Cd), nickel (Ni), and mercury (Hg) [6]. The sources for such metals are pesticides, metal complex dyes, fertilizers, mordants, bleaching agents, fixing agents, mordants, etc. [7]. Developed countries are taking strict actions to limit the heavy metal content in the wastewater. For mercury, lead, arsenic, cadmium, chromium, nickel, copper, and zinc, the current maximum contamination levels in India are 0.00003, 0.006, 0.050, 0.01, 0.05, 0.20, 0.25, 0.80 [8]. Heavy metals can be removed using techniques such chemical precipitation, chemical oxidation, ion exchange, electrodialysis, ultrafiltration, and adsorption [9]. Adsorption is the most reliable technique among these due to the limitations, such as low efficiency, high concentration of sludge, costly disposal, and sensitive operating conditions in the other mentioned techniques.

Domestic and industrial wastewater contains a huge amount of organic matter, water, nutrients (sulfur, phosphorous, nitrogen, etc.), and energy that can be recovered for a huge number of social, economic, and environmental purposes. However, due to the scattered and deficient data about the wastewater flows, the number of beneficial materials obtained from these streams is very low. This chapter provides a well-defined review of the sludge and wastewater flows and also highlights the economic importance of the matter contained in them. This chapter also includes social and environmental benefits that can be acquired by the treatment of wastewater [1].

2.2 Different Wastewater Treatments

On the surface of the earth, water appears to be abundant. However, it is a very precious and rare commodity and only a very small part of it is available for human consumption (almost 0.03%). The global population is rapidly growing, resulting in a high demand for water, yet the supply of available water remains steady. As a result, due to the lowest capacity of self-purification, its use must be minimized, and various procedures must be taken to return it to the environment with a minimum impurity burden [10]. During the previous two decades, environmental actions on a broad scale have occurred in the United States and Europe, resulting in tight environmental laws on industrial emissions. Effective cleaning methods and solutions must be purchased. Furthermore, many chemical companies have developed effluent treatment systems to comply with recently enacted rules in the nations where they do business.

The chemical industry is made up of enterprises that make industrial chemicals. Polymers, pharmaceutical goods, bulk intermediates and petrochemicals, basic industrials and other derivatives, fertilizers, and organic and inorganic chemicals are all examples of commodity chemicals, often known as basic chemicals. Chemical industry wastewaters frequently contain mutagenic, poisonous, carcinogenic, or nonbiodegradable organic and inorganic matter in various amounts, implying that many of the compounds in the effluent are difficult to breakdown. The petroleum hydrocarbons and surfactants liberated by the chemical industries degrade the efficiency of many treatment unit operations [11].

2.3 Technologies to Treat Chemical Industry Effluents

There are four methods of wastewater treatment [12].

Preliminary treatment: In this method, solid and macro materials are removed from the wastewater.
Primary treatment: Physical techniques are used to remove organic and inorganic materials, and the effluent produced during this step is referred to as primary effluent.
Secondary treatment: In this step, residual and suspended organic and other compounds are broken down which is brought about by bacterial action
Tertiary treatment: It involves residual disinfection and is usually a chemical process.

2.4 Oil–Water Separator—Treatment of Oily Effluent

The most common pollutant in the chemical industry is oil and grease (O&G). The effluents from petrochemical plants, oil refineries, chemical plants, food, and textile industries contain a very high level of oil and grease of the order of 200,000 mg/l [13, 14]. Free, dispersed, or emulsified oil and grease can be found in wastewater, with droplet sizes ranging from 150 mm to 20–150 mm and > 20 mm, respectively.

Gravity separation and skimming, deemulsification, dissolved air floatation, coagulation, and flocculation are all traditional procedures for treating oily wastewaters. Gravity separation and skimming are used to separate free oil from wastewater. The API separator, which is used in the main treatment phase to separate suspended particles and oil from wastewater effluents, is the most effective and low-cost oil-water separator. Separators of this type are built to the American Petroleum Institute's specifications. As a result, these separators are referred to as API separators.

However, emulsions and smaller oil droplets cannot be separated using the API separator. The process of sedimentation in a primary clarifier can be used to remove the oil particles adhering to the solid surface. The buoyancy of smaller oil droplets can be increased to enhance separation by the process of dissolved air floatation (DAF). To remove the emulsified oil, the DAF influent is deemulsified with thermal energy, chemicals, or both to promote coagulation and increase flock size. The emulsion is destabilized first by chemically treating the wastewater, and subsequently gravity separation is used. To stabilise the oil phase, heat is applied to the wastewater to lower viscosity, enhance density differences, and weaken interfacial coatings. The negative charge on the droplets is then neutralized by acidification and the addition of cationic polymer or alum. Following this phase, the pH is elevated to the alkaline area, allowing flock formation to occur. After that, the flock is separated from the adsorbed oil, and the sludge is thickened and dewatered [12].

2.5 Coagulation–Flocculation

Typically, wastewater treatment is accomplished through the sedimentation process, also known as clarification, in which the velocity of water is reduced to the point where the resulting velocity is less than the suspension velocity, causing the suspended particles to settle to the bottom due to gravitational force. The settled solids and suspended particles are next separated into sludge and scum. The effluent then travels through an effluent weir before proceeding to the next stage of treatment. Temperature, retention time, equipment quality, and tank design all influence the process's performance or efficacy. If sedimentation is done without coagulation or flocculation, only coarse suspended particles can be removed, which can even settle out without any chemical addition. This process is usually done in a clarification or sedimentation tank, a reservoir, etc. at the start of the process of treatment. The coagulation-flocculation technique involves the addition of organic or inorganic coagulants in the clarification tanks, which accelerates the sedimentation process. Aluminum hydroxide, aluminium sulphate, or high molecular weight cationic polymers may be employed as coagulants. At this stage of treatment, the inclusion of all of these chemicals has the goal of removing 90% of the particles from the wastewater.

2.6 Techniques for Treating Wastewater Using Adsorption

Adsorption is a natural process that involves the gathering of dissolved substances on the surface of a solid adsorbent. When the attractive forces of the liquid become less than the attractive forces at the carbon surface, adsorption occurs.

Natural and synthetic adsorbents have been categorized based on their origin. Natural adsorbents, such as clay, charcoal, clay minerals, ores, and zeolites, can be changed to improve their adsorption properties because they are very inexpensive and plentiful. Household wastes, sludge, synthetic wastes, and other materials can be used to make synthetic adsorbents. Adsorbents are characterized by good porosity, the nature of their adsorbing surface, and their pore structure. Sawdust, rice husk, coconut shells, and other waste materials can be used.

Organic pollutants can be adsorbed on activated carbon, but their adsorption is a complicated process and involves considerable difficulty. Dispersive, electrostatic, and chemical interactions, intrinsic properties of the solute and adsorbent, such as ionization or solubility constant and pore size distribution, respectively, the temperature of the system, solution properties, such as pH, and other factors all contribute to these difficulties [15]. Organic and inorganic impurities, bad odor, and taste can be removed from the wastewater by using powdered/granulated activated carbon. They have a large surface area, microporous structure, and nonpolar character, making them economically viable. 95% of the mass of activated carbon is composed of carbon and the rest 5% contains heteroatoms, such as nitrogen, oxygen, hydrogen, sulfur, and oxygen, which become associated with the activated carbon during the extraction from the raw material or during the activation and preparation procedures [16, 17]. Using bentonite as an adsorbent, Putra *et al.* [18] removed pharmaceutical effluents and antibiotics.

2.7 Adsorption of Dyes

Dyes can be removed by the process of adsorption. Large quantities of dyes are utilized by paper, pulp, and textile industries, which are then released into the bodies of water, giving rise to large amounts of toxic and carcinogenic colored wastewater, putting aquatic organisms in grave danger. The majority of the colours used are nonbiodegradable and resistant to aerobic digestion [19], posing a danger to the aquatic ecosystem. For the removal of dyes from wastewater, different low-cost adsorbents were utilized, such as activated rice husk [20] for colour removal, cedar sawdust, and broken brick for the removal of basic dye, methylene blue, respectively, with maximum adsorption reported to be 60 and 40 mg/L for each [21]. Wood shaving bottom ash was used to remove the azo reactive and red reactive 141 dyes (WBA). To improve the adsorption capability of the WBA/H2SO4 and WBA/H2O adsorbents, WBA was treated with 0.1 M H2SO4 and water, respectively. The presence of dissolved metals, pH of the solution, elution time, etc. affected the adsorption. The maximum dye absorption seen at 30°C was 41.5, 24.3, and 29.9 mg/L. The Langmuir model was used for the calculation [22]. Methylene blue dye has been removed using beer-brewery waste as a low-cost adsorbent. The list of various dyes and the adsorbents employed for their removal has been given in Table 2.1 below:

Table 2.1 List of few adsorbents and corresponding dyes removed [35].

Dye	Adsorbents used	Reference
Methylene blue	Activated carbon made from bituminous coal, activated carbon made from coal, activated carbon made from coal (KOH washed), activated carbon made from coal from Posidonia oceanic (L.) dead leaves, stalk-based activated carbon from cotton, activated Salix psammophila carbon, oil palm wood-based activated carbon, oil palm shell-based activated carbon.	[23–30]
Acid Blue 25	Activated carbon derived from waste tea	[31]
Everzol Red 3BS	Sepiolite, zeolite	[32, 33]
Acid Red 114	Nanoporous silica SBA-3	[34]

2.8 Zeolite in Wastewater Treatment

Zeolites, which are composed of hydrated aluminosilicate minerals with interconnected tetrahedral alumina and silica moieties, are the greatest adsorbents for removing cadmium [36]. Cadmium zeolites are useful for cadmium sequestration because they have a large surface area, superior ion exchange characteristics, and a hydrophilic nature. When compared to native zeolite, the modified zeolite's absorption capacity is greater. The alteration of a zeolite can be done in a variety of ways. Nanosized zeolites are more suited for heavy metal removal due to the presence of more accessible pores. The NaX nanozeolite (molar ratio 5.5 Na_2O: 1.0 Al_2O: 4.0 SiO_2: 190 H_2O) is a commonly used nanozeolite for cadmium removal from wastewater [37–41]. The microwave approach was used to synthesis NaX nanozeolite, and then the electrospinning method was utilized to make polyvinyl acetate polymer/NaX nanocomposite nanofibers, which were subsequently used to remove cadmium from wastewater [41]. At an ideal pH of 5, the maximum adsorption capacity was reported to be 838.7 mg/g, with 80% metal removal. This zeolite was functionally changed by cations, such as magnesium, sodium, potassium, or calcium replacing Al (III) and Si (IV) sites in the lattice, resulting in a net negative charge [36]. The Mg-modified zeolite has a number of advantages, including low cost, non-toxicity, abundance, and a bigger pore size of 40 to 50 nm with a cadmium removal capability of 98% at a pH of 7. Mg-modified zeolite has 1.5 times the adsorption capacity of Na/K-zeolite and 1.5 to 2.0 times the adsorption capacity of natural zeolite. Figure 2.1 shows the framework projections and the ring size for the most studied frameworks.

Different zeolites can be synthesized with the proper specific framework and varying chemical composition. More than 100 topologies of different zeolites have been reported, among which four are commonly used, zeolite Y, zeolite A, zeolite L, and ZSM-5. Zeolites have found several applications in separations and catalysis due to their ion-exchange properties, as well as the reaction and adsorption of molecules within their cages. Zeolites have numerous applications in the development of sensors and also in the improvement of existing sensing instruments [42–45].

Coal, used as a fuel in many industries, produces a by-product known as fly ash, which causes disposal problems and air pollution. Hydrothermal processes can be employed for zeolite formation from fly ash, which has a low-cost value [46]. At an optimal pH of 5, the fly ash was transformed into an amorphous aluminosilicate adsorbent with a maximum adsorption capacity of 24.246 mg/g and a cadmium removal of 80% [47]. Similarly,

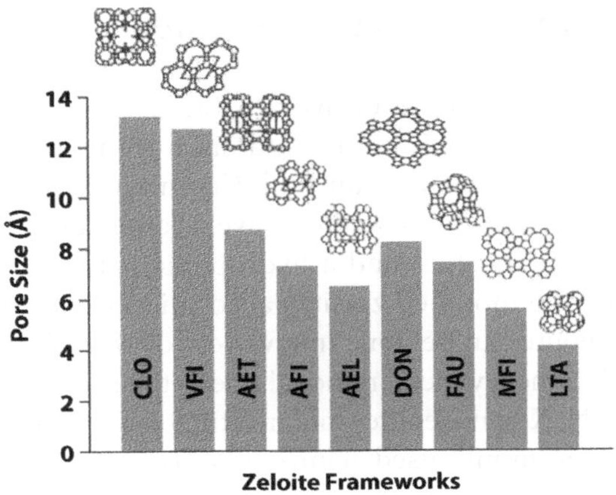

Figure 2.1 The framework projections and the ring size for the most-studied frameworks.

sodium hydroxide was utilized in a hydrothermal procedure to transform fly ash into zeolite adsorbent. This product is rich in micropores and has a large surface area, according to researchers, with an 80% cadmium removal capacity at an optimal pH of 7 to 8.

2.9 Negative Impact of Heavy Metals on Health

Metals occur naturally in the ecosystem. They readily lose electrons and form positively charged ions, so they are very good conductors of electricity. They are widely distributed over the whole globe, including the earth's atmosphere, water bodies, and crust, and can accumulate in the bodies of plants and animals. Among the 35 naturally occurring metals, 23 have an atomic mass greater than 40.04 and a high specific density greater than 5 gm/cm^3 and are thus classified as heavy metals, which have negative effects on organisms, as well as the ecosystem. These include tellurium, antimony, bismuth, thallium, tin, arsenic, gold, cerium, cadmium, gallium, cobalt, chromium, iron, copper, mercury, lead, nickel, manganese, silver, platinum, vanadium, uranium, and zinc [48–50]. Some of these metals, such as chromium, cobalt, magnesium, copper, iron, manganese, molybdenum, selenium, zinc, and nickel, are essential for various biochemical and physiological functions and are needed in a limited amount, the higher amount of which can cause chronic/acute toxicities [51]. Heavy metals can accumulate in the environment due to volcano eruptions, erosion, spring waters, bacterial, and human activities [52]. These metals get accumulated in the body and are transported to cells and tissues where they bind with the nucleic acids and proteins, disrupting their structure and causing serious consequences, such as brain damage, lung damage, kidney failure, liver damage, and several other diseases. Accumulation of heavy metals for long-time results in muscular, physical, and neurological degeneration, DNA damage, mutation, damage to the reproductive system, endocrine system, Alzheimer's disease, and Parkinson's disease [53, 54].

2.9.1 Origin of Heavy Metal Exposure to Humans

Toxic heavy metal ions are present widely in the environment, including the atmosphere, hydrosphere, lithosphere, and biosphere [55]. Humans are exposed to heavy metals through various anthropogenic activities, such as mining. These metals are present in the form of sulfide ores, oxides, etc. Lead, zinc, iron, gold, cobalt, and silver exist as sulfides, whereas manganese, aluminum, selenium, and gold exist in the form of oxides.

Some heavy metal sulfides may contain more than one metal, e.g., chalcopyrite (CuFeS2), which contains iron, as well as copper. These metals enter into the atmosphere, water, and soil through the process of mining, and then they come into contact with the bodies of organisms, including humans, and get accumulated in their bodies. Below are some of the metals and their sources, as well as the effects they cause on health.

Trace metals are required in minute quantities for the proper growth, development, and physiology of an organism. Trace metals are constantly cycled through living organisms as a result of their metabolic activities, and they are replenished by environmental absorption. Natural weathering processes on land, atmospheric deposition (for example, through rainfall), and waste streams absorbing water are the main pathways of trace metal delivery to the aquatic environment.

Table 2.2 Drinking water standards.

Metal ions	USPS (mg/l)	ISI (mg/l)	WHO (mg/l)	EPA (mg/l)
Pb^{2+}	0.010	0.010	0.100	0.050
Cd^{2+}	0.010	0.010	0.010	0.010
Zn^{2+}	5.000	5.000	-	5.000
As^{3+}	0.025	-	0.010	0.050

Metals adsorb on particulate matter and are then transferred to sediments in coastal seas. Table 2.2 shows the permissible limits for these ions in drinking water proposed by various international and national bodies (USPH 1989; ISI 1982; EPA 1975; WHO 1988). Consumers, however, rarely follow these guidelines [95–98].

2.9.1.1 *Arsenic*

Arsenate and arsenite compounds are the inorganic forms of arsenic and are harmful to most organisms on earth, including humans as well. Humans get exposed to this element through various industries, such as microelectronic and smelting industries, contaminated water, which contains arsenic washed into it from pesticides, wood preservatives, herbicides, paints, and fungicides [56].

Arsenic induces carcinogenesis by inducing epigenetic alterations, damaging the DNA maintenance system, and generating reactive oxygen species (ROS) [57, 58].

Some of the epigenetic changes caused by arsenic are DNA methylation, alteration of histones, etc., which are responsible for various types of malignancies [58]. Arsenic inhibits the expression of p21 protein due to the modification in the expression of p53 protein [59]. Arsenic promotes genotoxicity in mouse leucocytes and also in humans [60]. The effect of methylated arsenic was also studied, and it generated ROS in the spleen and liver and inhibited DNA repair processes, thereby increasing carcinogenesis [61]. Arsenic binds with the methyl-transferase, which causes the inhibition of the tumor-suppressor gene-coded DNA [62]. Arsenic poisoning causes cancers of the Kupffer cell, prostate, skin, and liver.

The toxicity caused by arsenic exposure may be acute, as well as chronic. Acute arsenic poisoning destroys gastrointestinal tissues and blood vessels and also affects the brain and heart. Chronic arsenic toxicity is known as arsenicosis, which causes skin problems, such as keratosis and pigmentation and can also damage the vital organs, thereby causing the death of an individual [63, 64]. Lower-level exposure can damage blood vessels and lead to the reduced production of leukocytes and erythrocytes, nausea, vomiting, pricking sensations in hands and legs, and can cause abnormal heartbeat. Long-term exposure can cause pulmonary diseases, peripheral vascular disease, neurological problems, the formation of skin lesions, hypertension, diabetes mellitus, and cardiovascular diseases [64]. In the United States, scientists have established that the amount of arsenic in drinking water should not be more than 50 μg/liter and the amount of arsenic should be lower in both food, as well as water.

2.9.1.2 Lead

Lead is an element with bright silver and a slightly bluish color. Food, cigarettes, drinking water, and domestic and industrial sources are the main sources of lead exposure. Gasoline, sanitation pipes, storage batteries, toys, vehicle exhausts, etc. are the industrial sources of lead that get released into the atmosphere and may get accumulated in the soil, which is taken up by plants and then goes to humans through food or drinking water [65, 66].

Lead toxicity causes the generation of ROS, which results in DNA damage, disruption of the DNA repair system, cellular tumor regulatory genes, and alteration of chromosomal structure and sequence [67]. Lead poisoning also causes problems in the central nervous system and gastrointestinal tract in humans [68]. Acute lead poisoning can cause indigestion, stomach aches, headaches, sleeplessness, tiredness, vertigo, confusion, hypertension, kidney failure, osteoarthritis, while chronic lead poisoning causes mental retardation, neonatal defects, brain damage, allergies, neurosis, weight loss, hyperactivity, other dreadful diseases, and even death [63].

2.9.1.3 Mercury

It is an odorless, shiny silver-white liquid metal that changes upon heating into a gas with no color and no odor. It has been employed in the manufacture of thermometers, dental amalgams, and different equipment in the metal and electrical industries. It can be inhaled when present in gaseous form. Humans get exposed to this metal through different anthropogenic activities, such as wastewater discharge from municipalities, industries, agriculture, mining, and incineration [69].

Mercury poisoning causes the generation of ROS that disrupts the biomolecules and causes an imbalance of antioxidant concentration in the body, thereby inducing malignancies [70].

Mercury combines with other elements and forms inorganic as well as organic forms of metal. Increased levels of organometallic mercury can cause renal failure and brain damage [71]. The organic form of the element is lipophilic and can, therefore, permeate the cell membranes. Long-term exposure to metallic mercury can cause decreased brain function, shyness, tremors, memory problems, irritability, and alteration in vision and hearing. Exposure to mercury for a short period can lead to vomiting, allergic skin, organ failure, diarrhea, hypertension, headache, tremors, hair loss, confusion, etc. Recent studies have also confirmed that if expecting mothers get exposed to the metal, it can cause loss of memory, reduction of motor neuron function, impaired speech, and malformation in the offspring [72]. Chronic mercury toxicity can cause a disease known as erethism, which is characterized by seizures, loss of memory, tremors, excitability, insomnia, and timidity.

2.10 Wastewater Treatment Using Different Zeolites

2.10.1 Natural Zeolites

Natural zeolites are aluminosilicate compounds that are hydrated have a high-sorption capacity and ion exchange characteristics, as well as a wide range of economic and environmental acceptance. The physicochemical properties determine the effectiveness of natural zeolites. The applications of natural zeolites are due to their porous 3-D structure, and the isomorphic replacement of silicon by aluminium in the main structure. Natural zeolites are categorized as cationic exchangers because they contain a significant negative charge on their surfaces. Zeolites can, therefore, be used to remove metal cations from wastewater. However, zeolites can be chemically changed by adsorbing organic surfactants or inorganic salts, resulting in surfactant micelles or positively charged oxyhydroxides that can be employed for anion exchange [73].

Natural zeolites release nonhazardous cations, such as K^+ and Mg^{2+} into the atmosphere, making them suitable candidates for cation exchange. Moreover, they are cheap and have a compact size. The natural or modified zeolites take advantage of the type and quantity of adsorbent employed, as well as the particle size distribution of the adsorbent, pH value of the solution, pressure, temperature, the presence of other anions or organic compounds, etc. in the wastewater treatment. Normally, natural zeolites are utilized to treat wastewater via the column exchange method [73, 74].

For a long time, natural zeolites have been widely used in the wastewater treatment industry. Heavy metals (such as Cr, Zn, Cd, Pb, Mn, Cu, Fe, and others) pose a major hazard to the ecosystem, their removal is necessary. This can be achieved by combining natural zeolites with ion exchange, membrane filtration, flotation, and other techniques, agglomeration, and oxidation-reduction methods [75]. The usage of natural zeolites in wastewater treatment and how they might be changed to have high efficacy have been researched by researchers, which can be brought about by ion exchange, acid treatment, or surfactant functionalization. These zeolites adsorb anions and organic matter at a high rate [76].

2.11 Treatment of Surface Waters, Ground, and Underground Waters

Natural zeolites can also be used to remove humic acid and ammonia from surface waters, with the amount of ammonia and humic acid removed depending on the initial concentrations, pH, contact time, and temperature. The results showed that natural zeolites perform best at pH levels similar to those seen in natural waterways [77]. Moreover, natural and modified zeolites can be used for the removal of Mn and Fe ions from underground waters. The Mn and Fe removal levels are 61% to 100% and 22% to 90%, respectively, for natural zeolite-clinoptilolite [78]. The many formulations for removing arsenic from groundwater are far more important and require further research.

2.12 Drinking and Greywater Treatment

The removal of pollutants from drinking water can be brought about by many conventional methods, such as ion exchange, membrane processes, coagulation followed by filtration. Adsorption techniques have proven to be cost-effective, effective, and simple to develop and use. The removal of Mn, Cu, and Zn was carried out experimentally using natural clinoptilolite zeolite and clinoptilolite combined with Fe. Natural zeolites have excellent efficiency due to their selectivity towards cations. Clinoptilolite-Fe zeolite is an excellent adsorbent that is also inexpensive, safe for the environment and humans, and easily regenerated and synthesized [79]. The removal of

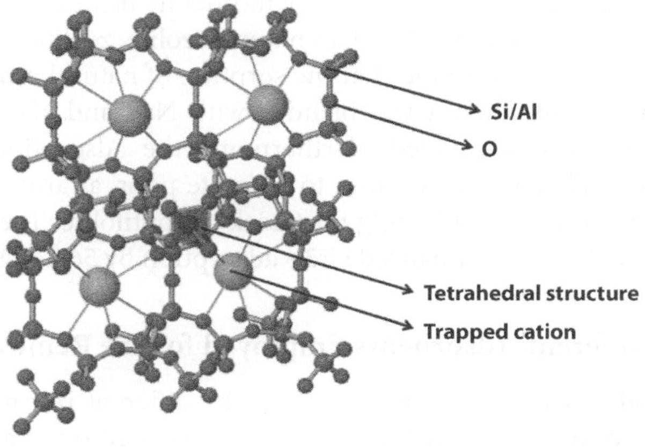

Figure 2.2 The tetrahedral framework of clinoptilolite zeolite. Modified from IZA [83].

arsenic from water can be brought about by using modified natural zeolites prepared from different iron solutions and also aluminum/iron hydroxide. The performance can be enhanced by using granular sorbents, which accelerate the throughput and rate of the process. The clinoptilolite-Fe zeolite could adsorb more than 90 mg of arsenic per kg of the zeolite [80, 81]. By using clinoptilolite along with FeO, $Cu2^{+}$ cations can be removed from the wastewater. Modified clinoptilolite adsorbed 37.5 mg/g Cu ions, while the natural clinoptilolite adsorbed 13.6 mg/g Cu ions [82]. The tetrahedral structure of clinoptilolite is given in Figure 2.2.

2.13 Heavy Metal Removal Comparison by Zeolites

It is a very difficult process to immobilize heavy metals, and their removal depends on the type of adsorbent. Using zeolites, whether natural or manufactured, researchers are attempting to remove heavy metals from wastewater. The S/L ratio, or the ratio of zeolite to wastewater, as well as contact time, temperature, solution pH, and initial metal ion concentration, all influence the ratio of metals adsorbed on the surface of zeolites. The removal of various metal ions using various types of adsorbents, as well as the best adsorption test settings, is summarized in the table below.

2.13.1 Different Adsorbents Used to Remove Cr^{3+}

The heavy metal ion Cr^{3+} may be removed using a modified zeolite, as shown in Table 2.3, and the adsorption maximum can reach 83.2 mg/g, as shown in Table 2.3. The removal of Cr^{3+} was also brought about by using zeolites from kaolin, natural mordenite, mordenite blended with NaP and NaY, Greece natural zeolite, Brazilian natural zeolite, zeolite NaP1 and 4A from coal fly ash. It was found that the sorption of natural mordenite was only 3.5 mg/g, and when it was blended with NaP and NaY, 83.2 mg/g of adsorption was discovered. Furthermore, the adsorption ability of Brazilian natural zeolite was found to be quite poor, at around 3 to 14.5 mg/g, compared to only 4.1 mg/g for Greek clinoptilolite. The use of coal fly zeolite NaP1 and 4A enhanced Cr3+ adsorption by 56.4 mg/g [84–88].

2.13.2 Different Adsorbents Employed for the Removal of Cd^{3+}

The removal of Cd^{2+} has been carried out by different researchers using different types of zeolites, natural as well as modified ones, and the process has been summarized [89, 90] in Table 2.4. The adsorption of Cd^{2+} by

Table 2.3 Different zeolites compared for their ability to remove Cr^{3+} [84–87].

Zeolite	Natural mordenite	NaP	Blend of NaX and NaA	Blend of NaY and NaP	Zeolite 4A	Hydroxysodalite	Natural calcite	NaP1
Zeolite origin	From Kaolin	From mordenite	From mordenite	From mordenite	From CFA	From kaolin	Brazil	From CFA
Metal ion	Cr^{3+}	Cr^{3+}	Cr^{3+}	Cr^{3+}	Cr^{3+}	Cr^{3+}	Cr^{3+}	Cr^{3+}
T (◦C)	25	25	25	25	25	25	55	22
Metal concentration (mg/L)	500	500	500	500	50–100	500	50–250	100
Time (h)	24	24	24	24	1	24	1	6
pH					3–4		6	4
S/L ratios (g/mL)	1/200	1/200	1/200	1/200	1/1000	1/200	1/60	1/100
Metal sorption (mg/g)	3.5	52	71.1	83.2	38.7–56.4	34.7	3.0–14.5	43.6

Table 2.4 Comparison for Cd^{2+} removal by different zeolites [89–93].

Zeolite	Natural scolecite	Bentonite	Zeolite A	Zeolite X	Clinoptilolite	Clinoptilolite	NaP1	NaX + activated carbon
Zeolite origin	Brazil	Commercial	From kaolin	From kaolin	Greece	Croatia	From CFA	From CFA
Metal ion	Cd^{2+}	Cd^{2+}	Cd^{2+}	Cd^{2+}	Cd^{2+}	Cd^{2+}	Cd^{2+}	Cd^{2+}
Time (h)	1	1.5	0.5	0.5	1	24	1	24
T (°C)	55	30	25	25	22	70	22	25
Metal conc (mg/L)	50–250	25	20	20	100	1024	100	500
pH	6	6	7.5	7.5	6	4.5	6	
S/L ratio (g/mL)	1/60	1/40	1/125	1/125	1/100	1/100	1/100	1/500
Metal sorption (mg/g)	2.9–6.0	=9.44	71	92	4.1	13.5	50.8	129.3

Croatian natural clinoptilolite is 13.5 mg/g in wastewater where the Cd^{2+} concentration is 1,024 mg/L, which is very high compared to the removed quantity by the zeolite. When the metal ion content is less than 100 mg/L, the adsorption effectiveness of unmodified zeolite is reduced [86, 91, 92]. When the starting concentration of Cd2+ in the wastewater is 500 mg/L, the adsorption of Cd2+ utilizing synthesized zeolites from CFA is higher and recorded at 129.3 mg/g [93]. The adsorption of Cd2+ increased 10 times when natural zeolites were replaced by synthetic ones.

2.13.3 Removal of Cu^{2+} by Different Adsorbents

Cu^{2+} can be removed from waste using clinoptilolite from Bulgaria and Greece, but the adsorption is less than 5.3 mg/g, which is very little [86, 94, 95]. Synthetic or modified zeolites can be used to remove copper ions from water, and their adsorption is greater as compared to natural zeolites. The adsorption maximum can reach 101.7 mg/g when NaX, blended with carbon, is used as an adsorbent [94]. The Bulgarian natural zeolite modified with NaCl, CH_3COONa and NaOH can be used to adsorb copper ions from wastewater with a Cu^{2+} ion content of about 50 mg/L and the adsorption is approximately 4.8 to 4.9 mg/g, which is higher than that of natural zeolite, which is approximately 2.9 mg/g when all other circumstances are held constant.

2.13.4 Different Adsorbents Used to Remove Pb^{2+}

The adsorption of Pb^{2+} ions shown by natural zeolite of Croatia is 10 mg/g higher than that of Turkey zeolite (78.7 mg/g at pH 4.5, temperature 70°C and sorption time 24 hours) [89, 96] which indicates that for the removal of lead from wastewater, the structural diversity is very significant. When Croatia natural zeolite was treated with 2M NaCl solution at 70°C for 24 hours while leaving the other testing conditions the same, the Pb sorption quantity increased to 91.2 mg/g. Different types of zeolites used for the adsorption of Pb^{2+} ions have been given in Table 2.5.

2.13.5 Removal of Zn^{2+} by Different Adsorbents

Zn^{2+} ions can be removed from wastewater by using natural, modified, and synthesized zeolites as shown in Table 2.6. When the Zn starting concentration is 100 mg/L and the sorption time is 2 hours, natural clinoptilolite from Greece adsorbs 3.1 mg/g Zn^{2+} ions, while synthetic zeolites from CFA adsorb 40.4 mg/g Zn^{2+} ions [86, 87].

Table 2.5 Different zeolites used for the removal of Pb^{2+} [89, 96].

Zeolite	Zeolite A	Zeolite X	Clinoptilolite	Clinoptilolite	Na-clinoptilolite	NaX+ activated carbon
Zeolite origin	From kaolin	From kaolin	Croatia	Turkey	Croatia	CFA
Metal ion	Pb^{2+}	Pb^{2+}	Pb^{2+}	Pb^{2+}	Pb^{2+}	Pb^{2+}
Time (h)	0.5	0.5	24	2	24	24
T (°C)	25	25	70	25	70	25
Metal conc (mg/L)	20	20	100	10–100	100	1000
pH	7.5	7.5	4.5	2–7	4.5	
S/L ratio (g/mL)	1/250	1/125	1/100	1/100	1/100	1/500
Metal sorption (mg/g)	213	187	78.7	10	91.2	228

Table 2.6 Different zeolites used for the removal of Zn^{2+} [86, 87].

Zeolite	**Zeolite A**	**Zeolite 4A**	**Zeolite X**	**NaP1**	**Clinoptilolite**	**Clinoptilolite**
Zeolite origin	From kaolin	From CFA	From kaolin	From CFA	Greece	Turkey
Metal ion	Zn^{2+}	Zn^{2+}	Zn^{2+}	Zn^{2+}	Zn^{2+}	Zn^{2+}
Time (h)	0.5	2	0.5	1	1	5.5
T (°C)	25	25	0.5	22	22	25
Metal conc (mg/L)	20	50–100	20	100	100	25
pH	7.5	3–4	7.5	6	6	6–7
S/L ratio (g/mL)	1/125	1/1000	1/125	1/100	1/100	1/50
Metal sorption (mg/g)	28.6	9.6–40.4	41	32.6	3.1	22

2.14 Adsorption Kinetics and Thermodynamics

Natural, modified, and synthetic zeolites, as well as modified and synthetic zeolites, can all be utilized to remove heavy metals from wastewater, according to the findings of the literature review. In diverse ion-exchange processes, zeolite samples from distinct areas have variable sorption behaviour. As a result, researchers investigated the sorption characteristics of natural and synthetic zeolites for heavy metals in order to see if these zeolites might be used to cleanse real metal wastewaters by looking at the heavy metals' kinetics and thermodynamics. On porous adsorbents, adsorption occurs in three stages: (1) solute transfer from the bulk solution to the sorbent's external surface via a liquid boundary layer (film resistance); (2) solute transfer to the intraparticle active site via the sorbent surface; and (3) solute transfer to the intraparticle active site via the sorbent surface. The amount of solute adsorbed and the rate at which it is adsorbed are determined by one or more of the previously described processes. Liquid-phase sorption is a pseudo-first-order reaction that is governed by the solid capacity, which directly connects the rate of change of sorbate uptake over time to the difference in saturation concentration and the quantity of solid uptake over time. The Lagergren equation is the most commonly utilized equation in a liquid/solid system. The Lagergren pseudo-second-order model was shown to be the best fit for the spontaneity of Cr ion adsorption by natural scolecite from Brazil, and thermodynamic data revealed that the reaction is an endothermic cation-exchange mechanism [85]. With increasing temperature, the rate of sorption increased. The adsorption of Cr ions on zeolite 4A made from coal fly ash was examined and shown to follow pseudo-second-order kinetics and be nearly instantaneous [86, 87].

Cadmium was adsorbed on Brazilian natural scolecite, with pseudo-second-order kinetics [85]. Kardjali natural zeolite was likewise able to absorb Cd^{2+}, and the process was second-order irreversible [91]. The Pb ion was adsorbed on a composite of NaX, activated carbon from CFA, and Turkish clinoptilolite [96], with pseudo-second-order kinetics. The rate of sorption increased with the increase in temperature. The adsorption of Cr ions was studied on zeolite 4A prepared from coal fly ash and followed pseudo-second-order kinetics and was practically instantaneous [86, 87].

2.15 Conclusion

To remove contaminants from wastewaters, traditional technologies, such as ion exchange, membrane processes, and coagulation followed by filtration are now in use. Adsorption methods have shown to be the most cost-effective, efficient, and simple to use in all of them. Heavy metal ion removal from wastewater is difficult and is dependent on the adsorbent used. Using zeolites, both natural and manufactured, researchers are attempting to extract heavy metals from wastewater. The chapter demonstrated that both natural and synthetic zeolites can be employed to sorb hazardous metal ions successfully. In addition, the Si/Al ratio, contact time, temperature, solution pH, and initial metal ion concentration all affect zeolites' efficiency.

References

1. Asano, T., Burton, F., Leverenz, H., *Water Reuse: Issues, Technologies, and Applications*, McGraw-Hill Education, Hongkong, 2007.
2. Mateo-Sagasta, J., Raschid-Sally, L., Thebo, A., Global wastewater and sludge production, treatment, and use, in: *Wastewater*, pp. 15–38, Springer, Dordrecht, 2015.
3. MacCarthy, P., Klusman, R.W., Cowling, S.W., Rice, J.A., Water analysis. *Anal. Chem.*, 67, 12, 525–582, 1995.
4. Clement, R.E., Koester, C.J., Eiceman, G.A., Environmental analysis. *Anal. Chem.*, 65, 12, 85–116, 1993.
5. Renge, V.C., Khedkar, S.V., Pande, S.V., Removal of heavy metals from wastewater using low cost adsorbents: A review. *Sci. Rev. Chem. Commun.*, 2, 4, 580–584, 2012.
6. Mehdipour, S., Vatanpour, V., Kariminia, H.R., Influence of ion interaction on lead removal by a polyamide nanofiltration membrane. *Desalination*, 362, 84–92, 2015.
7. Rao, K.S., Mohapatra, M., Anand, S., Venkateswarlu, P., Review on cadmium removal from aqueous solutions. *Int. J. Eng. Sci. Technol.*, 2, 7, 81–103, 2010.
8. Gopalakrishnan, A., Krishnan, R., Thangavel, S., Venugopal, G., Kim, S.J., Removal of heavy metal ions from pharma-effluents using graphene-oxide nanosorbents and study of their adsorption kinetics. *J. Ind. Eng. Chem.*, 30, 14–19, 2015.
9. Fu, F. and Wang, Q., Removal of heavy metal ions from wastewaters: A review. *J. Environ. Manage.*, 92, 3, 407–418, 2011.
10. Gamper-Rabindran, S. and Finger, S.R., Does industry self-regulation reduce pollution? Responsible care in the chemical industry. *J. Regul. Econ.*, 43, 1, 1–30, 2013.
11. Srinivasan, P., Bosco, A.J., Kalaivizhi, R., Selvi, J.A., Sivakumar, P., Adsorption isotherm and kinetic study of direct orange 102 dyes on TNJ activated carbon. *Mater. Today Proc.*, 34, 389–394, 2021.
12. Awaleh, M.O. and Soubaneh, Y.D., Wastewater treatment in chemical industries: The concept and current technologies. *Hydrol. Curr. Res.*, 5, 1, 1, 2014.

13. Cheryan, M. and Rajagopalan, N., Membrane processing of oily streams. Wastewater treatment and waste reduction. *J. Membr. Sci.*, 151, 1, 13–28, 1998.
14. Hu, G., Li, J., Zeng, G., Recent development in the treatment of oily sludge from petroleum industry: A review. *J. Hazard. Mater.*, 261, 470–490, 2013.
15. Moreno-Castilla, C., Adsorption of organic molecules from aqueous solutions on carbon materials. *Carbon*, 42, 1, 83–94, 2004.
16. Amokrane, A., Comel, C., Veron, J., Landfill leachates pretreatment by coagulation-flocculation. *Water Res.*, 31, 11, 2775–2782, 1997.
17. Surmacz-Gorska, J., *Degradation of Organic Compounds in Municipal Landfill Leachate*, Publishers of Environmental Engineering Committee of Polish Academy of Sciences, Lublin, 2001.
18. Putra, E.K., Pranowo, R., Sunarso, J., Indraswati, N., Ismadji, S., Performance of activated carbon and bentonite for adsorption of amoxicillin from wastewater: Mechanisms, isotherms and kinetics. *Water Res.*, 43, 9, 2419–2430, 2009.
19. Ardejani, F.D., Badii, K., Limaee, N.Y., Mahmoodi, N.M., Arami, M., Shafaei, S.Z., Mirhabibi, A.R., Numerical modelling and laboratory studies on the removal of direct red 23 and direct red 80 dyes from textile effluents using orange peel, a low-cost adsorbent. *Dyes Pigments*, 73, 2, 178–185, 2007.
20. Gupta, V.K., Mittal, A., Jain, R., Mathur, M., Sikarwar, S., Adsorption of Safranin-T from wastewater using waste materials—Activated carbon and activated rice husks. *J. Colloid Interface Sci.*, 303, 1, 80–86, 2006.
21. Hamdaoui, O., Batch study of liquid-phase adsorption of methylene blue using cedar sawdust and crushed brick. *J. Hazard. Mater.*, 135, 1-3, 264–273, 2006.
22. Leechart, P., Nakbanpote, W., Thiravetyan, P., Application of 'waste' wood-shaving bottom ash for adsorption of azo reactive dye. *J. Environ. Manage.*, 90, 2, 912–920, 2009.
23. El Qada, E.N., Allen, S.J., Walker, G.M., Adsorption of methylene blue onto activated carbon produced from steam activated bituminous coal: a study of equilibrium adsorption isotherm. *Chem. Eng. J.*, 124, 1-3, 103–110, 2006.
24. Gong, G.Z., Qiang, X., Zheng, Y.F., Ye, S.F., Chen, Y.F., Regulation of pore size distribution in coal-based activated carbon. *New Carbon Mater.*, 24, 2, 141–146, 2009.
25. Dural, M.U., Cavas, L., Papageorgiou, S.K., Katsaros, F.K., Methylene blue adsorption on activated carbon prepared from posidonia oceanica (L.) dead leaves: Kinetics and equilibrium studies. *Chem. Eng. J.*, 168, 1, 77–85, 2011.
26. Girgis, B.S., Smith, E., Louis, M.M., El-Hendawy, A.N.A., Pilot production of activated carbon from cotton stalks using H3PO4. *J. Anal. Appl. Pyrolysis*, 86, 1, 180–184, 2009.
27. Bao, Y. and Zhang, G., Study of adsorption characteristics of methylene blue onto activated carbon made by salix psammophila. *Energy Procedia*, 16, 1141–1146, 2012.

28. Ahmad, A.L., Loh, M.M., Aziz, J.A., Preparation and characterization of activated carbon from oil palm wood and its evaluation on methylene blue adsorption. *Dyes Pigments*, 75, 2, 263–272, 2007.
29. Vargas, A.M., Cazetta, A.L., Kunita, M.H., Silva, T.L., Almeida, V.C., Adsorption of methylene blue on activated carbon produced from flamboyant pods (delonixregia): Study of adsorption isotherms and kinetic models. *Chem. Eng. J.*, 168, 2, 722–730, 2011.
30. Tan, I.A.W., Ahmad, A.L., Hameed, B.H., Enhancement of basic dye adsorption uptake from aqueous solutions using chemically modified oil palm shell activated carbon. *Colloids Surf. A Physicochem. Eng. Asp.*, 318, 1-3, 88–96, 2008.
31. Auta, M. and Hameed, B.H., Preparation of waste tea activated carbon using potassium acetate as an activating agent for adsorption of acid blue 25 dye. *Chem. Eng. J.*, 171, 2, 502–509, 2011.
32. Ozdemir, O., Armagan, B., Turan, M., Celik, M.S., Comparison of the adsorption characteristics of azo-reactive dyes on mesoporous minerals. *Dyes Pigments*, 62, 1, 49–60, 2004.
33. Mall, I.D., Srivastava, V.C., Agarwal, N.K., Removal of Orange-G and Methyl Violet dyes by adsorption onto bagasse fly ash—Kinetic study and equilibrium isotherm analyses. *Dyes Pigments*, 69, 3, 210–223, 2006.
34. Anbia, M. and Salehi, S., Removal of acid dyes from aqueous media by adsorption onto amino-functionalized nanoporous silica SBA-3. *Dyes Pigments*, 94, 1, 1–9, 2012.
35. Rashed, M.N., Adsorption technique for the removal of organic pollutants from water and wastewater, in: *Organic Pollutants-Monitoring, Risk and Treatment*, vol. 7, pp. 167–194, 2013.
36. Choi, H.J., Yu, S.W., Kim, K.H., Efficient use of mg-modified zeolite in the treatment of aqueous solution contaminated with heavy metal toxic ions. *J. Taiwan Inst. Chem. Eng.*, 63, 482–489, 2016.
37. Erdem, E., Karapinar, N., Donat, R., The removal of heavy metal cations by natural zeolites. *J. Colloid Interface Sci.*, 280, 2, 309–314, 2004.
38. Jha, V.K., Matsuda, M., Miyake, M., Sorption properties of the activated carbon-zeolite composite prepared from coal fly ash for Ni2+, Cu2+, Cd2+ and Pb2+. *J. Hazard. Mater.*, 160, 1, 148–153, 2008.
39. Ibrahim, H.S., Jamil, T.S., Hegazy, E.Z., Application of zeolite prepared from Egyptian kaolin for the removal of heavy metals: II. Isotherm models. *J. Hazard. Mater.*, 182, 1-3, 842–847, 2010.
40. Aliabadi, M., Irani, M., Ismaeili, J., Piri, H., Parnian, M.J., Electrospun nanofiber membrane of PEO/Chitosan for the adsorption of nickel, cadmium, lead and copper ions from aqueous solution. *Chem. Eng. J.*, 220, 237–243, 2013.
41. Rad, L.R., Momeni, A., Ghazani, B.F., Irani, M., Mahmoudi, M., Noghreh, B., Removal of Ni2+ and Cd2+ ions from aqueous solutions using electrospun PVA/zeolite nanofibrous adsorbent. *Chem. Eng. J.*, 256, 119–127, 2014.

42. Xu, X., Wang, J., Long, Y., Zeolite-based materials for gas sensors. *Sensors*, 6, 12, 1751–1764, 2006.
43. Fong, Y.Y., Abdullah, A.Z., Ahmad, A.L., Bhatia, S., Zeolite membrane based selective gas sensors for monitoring and control of gas emissions. *Sensor Lett.*, 5, 3-4, 485–499, 2007.
44. Sahner, K., Hagen, G., Schönauer, D., Reiß, S., Moos, R., Zeolites—Versatile materials for gas sensors. *Solid State Ionics*, 179, 40, 2416–2423, 2008.
45. Zheng, Y., Li, X., Dutta, P.K., Exploitation of unique properties of zeolites in the development of gas sensors. *Sensors*, 12, 4, 5170–5194, 2012.
46. Hui, K.S., Chao, C.Y.H., Kot, S.C., Removal of mixed heavy metal ions in wastewater by zeolite 4A and residual products from recycled coal fly ash. *J. Hazard. Mater.*, 127, 1-3, 89–101, 2005.
47. Javadian, H., Ghorbani, F., Tayebi, H.A., Asl, S.H., Study of the adsorption of cd (II) from aqueous solution using zeolite-based geopolymer, synthesized from coal fly ash; Kinetic, isotherm and thermodynamic studies. *Arab. J. Chem.*, 8, 6, 837–849, 2015.
48. Duffus, J.H., "Heavy metals" a meaningless term? (IUPAC technical report). *Pure Appl. Chem.*, 74, 5, 793–807, 2002.
49. Li, F., Qiu, Z., Zhang, J., Liu, W., Liu, C., Zeng, G., Investigation, pollution mapping and simulative leakage health risk assessment for heavy metals and metalloids in groundwater from a typical brownfield, middle China. *Int. J. Environ. Res. Public Health*, 14, 7, 768, 2017.
50. Bradl, H. (Ed.), *Heavy Metals in the Environment: Origin, Interaction and Remediation*, Elsevier, Germany, 2005.
51. Levander, O.A. and Whanger, P.D., Deliberations and evaluations of the approaches, endpoints and paradigms for selenium and iodine dietary recommendations. *J. Nutr.*, 126, suppl_9, 2427S–2434S, 1996.
52. Florea, A.-M., Dopp, E., Obe, G., Rettenmeier, A.W., H., Genotoxicity of organometallic species, in: *Organic Metal and Metalloid Species in the Environment: Analysis, Distribution, Processes and Toxicological Evaluation*, A.V. Hirner, and Emons, (Eds.), pp. 205–219, Springer-Verlag, Heidelberg, 2004.
53. Monisha, J., Tenzin, T., Naresh, A., Blessy, B.M., Krishnamurthy, N.B., Toxicity, mechanism and health effects of some heavy metals. *Interdiscip. Toxicol.*, 7, 2, 60–72, 2014.
54. Järup, L., Hazards of heavy metal contamination. *Br. Med. Bull.*, 68, 1, 167–182, 2003.
55. Krishna, A.K. and Mohan, K.R., Distribution, correlation, ecological and health risk assessment of heavy metal contamination in surface soils around an industrial area, Hyderabad, India. *Environ. Earth Sci.*, 75, 5, 411, 2016.
56. Sauvé, S., Time to revisit arsenic regulations: Comparing drinking water and rice. *BMC Public Health*, 14, 1, 1–5, 2014.
57. Martinez, V.D., Vucic, E.A., Becker-Santos, D.D., Gil, L., Lam, W.L., Arsenic exposure and the induction of human cancers. *J. Toxicol.*, 2011, 1–13, 2011.

58. Bjørklund, G., Aaseth, J., Chirumbolo, S., Urbina, M.A., Uddin, R., Effects of arsenic toxicity beyond epigenetic modifications. *Environ. Geochem. Health*, 40, 3, 955–965, 2018.
59. Park, Y.H., Kim, D., Dai, J., Zhang, Z., Human bronchial epithelial BEAS-2B cells, an appropriate *in vitro* model to study heavy metals induced carcinogenesis. *Toxicol. Appl. Pharmacol.*, 287, 3, 240–245, 2015.
60. Banu, B.S., Danadevi, K., Jamil, K., Ahuja, Y.R., Rao, K.V., Ishaq, M., *In vivo* genotoxic effect of arsenic trioxide in mice using comet assay. *Toxicology*, 162, 3, 171–177, 2001.
61. Hartwig, A. and Schwerdtle, T., Interactions by carcinogenic metal compounds with DNA repair processes: Toxicological implications. *Toxicol. Lett.*, 127, 1-3, 47–54, 2002.
62. García-Esquinas, E., Pollán, M., Umans, J.G., Francesconi, K.A., Goessler, W., Guallar, E., Navas–Acien, A., Arsenic exposure and cancer mortality in a US-based prospective cohort: The strong heart study. *Cancer Epidemiol. Biomarkers Prev.*, 22, 11, 1944–1953, 2013.
63. Martin, S. and Griswold, W., Human health effects of heavy metals, in: *Environmental Science and Technology Briefs for Citizens*, vol. 15, pp. 1–6, 2009.
64. Huy, T.B., Tuyet-Hanh, T.T., Johnston, R., Nguyen-Viet, H., Assessing health risk due to exposure to arsenic in drinking water in Hanam Province, Vietnam. *Int. J. Environ. Res. Public Health*, 11, 8, 7575–7591, 2014.
65. Thürmer, K., Williams, E., Reutt-Robey, J., Autocatalytic oxidation of lead crystallite surfaces. *Science*, 297, 5589, 2033–2035, 2002.
66. Wani, A.L., Ara, A., Usmani, J.A., Lead toxicity: A review. *Interdiscip. Toxicol.*, 8, 2, 55, 2015.
67. Silbergeld, E.K., Waalkes, M., Rice, J.M., Lead as a carcinogen: Experimental evidence and mechanisms of action. *Am. J. Ind. Med.*, 38, 3, 316–323, 2000.
68. Markowitz, M., Lead poisoning. *Pediatr. Rev.*, 21, 10, 327–335, 2000.
69. Rahimzadeh, M.R., Rahimzadeh, M.R., Kazemi, S., Moghadamnia, A.A., Cadmium toxicity and treatment: An update. *Caspian J. Intern. Med.*, 8, 3, 135, 2017.
70. Crespo-López, M.E., Macêdo, G.L., Pereira, S., II, Arrifano, G.P., Picanço-Diniz, D.L., do Nascimento, J.L.M., Herculano, A.M., Mercury and human genotoxicity: Critical considerations and possible molecular mechanisms. *Pharmacol. Res.*, 60, 4, 212–220, 2009.
71. Alina, M., Azrina, A., MohdYunus, A.S., MohdZakiuddin, S., MohdIzuan Effendi, H., Muhammad Rizal, R., Heavy metals (mercury, arsenic, cadmium, plumbum) in selected marine fish and shellfish along the straits of Malacca. *Int. Food Res. J.*, 19, 1, 135–140, 2012.
72. Bernhoft, R.A., Mercury toxicity and treatment: A review of the literature. *J. Environ. Public Health*, 2012, 460508, 1–10, 2012.
73. Margeta, K., Logar, N.Z., Siljeg, M., Farkas, A., Natural zeolites in water treatment–how effective is their use, in: *Water Treatment*, vol. 5, pp. 81–112, 2013.

74. Caputo, D. and Pepe, F., Experiments and data processing of ion exchange equilibria involving Italian natural zeolites: A review. *Microporous Mesoporous Mater.*, 105, 3, 222–231, 2007.
75. Fu, F. and Wang, Q., Removal of heavy metal ions from wastewaters: A review. *J. Environ. Manage.*, 92, 3, 407–418, 2011.
76. Wang, S. and Peng, Y., Natural zeolites as effective adsorbents in water and wastewater treatment. *Chem. Eng. J.*, 156, 1, 11–24, 2010.
77. Moussavi, G., Talebi, S., Farrokhi, M., Sabouti, R.M., The investigation of mechanism, kinetic and isotherm of ammonia and humic acid co-adsorption onto natural zeolite. *Chem. Eng. J.*, 171, 3, 1159–1169, 2011.
78. Inglezakis, V.J., Doula, M.K., Aggelatou, V., Zorpas, A.A., Removal of iron and manganese from underground water by use of natural minerals in batch mode treatment. *Desalination Water Treat.*, 18, 1-3, 341–346, 2010.
79. Doula, M.K., Simultaneous removal of Cu, Mn and Zn from drinking water with the use of clinoptilolite and its Fe-modified form. *Water Res.*, 43, 15, 3659–3672, 2009.
80. Li, Z., Jean, J.S., Jiang, W.T., Chang, P.H., Chen, C.J., Liao, L., Removal of arsenic from water using Fe-exchanged natural zeolite. *J. Hazard. Mater.*, 187, 1-3, 318–323, 2011.
81. Habuda-Stanic, M., Kalajdzic, B., Kules, M., Velic, N., Arsenite and arsenate sorption by hydrous ferric oxide/polymeric material. *Desalination*, 229, 1-3, 1–9, 2008.
82. Doula, M.K. and Dimirkou, A., Use of an iron-overexchanged clinoptilolite for the removal of Cu2+ ions from heavily contaminated drinking water samples. *J. Hazard. Mater.*, 151, 2-3, 738–745, 2008.
83. Perez-Botella, E., Misturini, A., Sala, A., Palomino, M., Corma, A., Sastre, G., Rey, F., Insights into adsorption of linear, monobranched, and dibranched alkanes on pure silica STW zeolite as a promising material for their separation. *J. Phys. Chem. C*, 124, 49, 26821–26829, 2020.
84. Covarrubias, C., García, R., Arriagada, R., Yanez, J., Garland, M.T., Cr (III) exchange on zeolites obtained from kaolin and natural mordenite. *Microporous Mesoporous Mater.*, 88, 1-3, 220–231, 2006.
85. Dal Bosco, S.M., Jimenez, R.S., Carvalho, W.A., Removal of toxic metals from wastewater by Brazilian natural scolecite. *J. Colloid Interface Sci.*, 281, 2, 424–431, 2005.
86. Alvarez-Ayuso, E., Garcıa-Sanchez, A., Querol, X., Purification of metal electroplating wastewaters using zeolites. *Water Res.*, 37, 20, 4855–4862, 2003.
87. Hui, K.S., Chao, C.Y.H., Kot, S.C., Removal of mixed heavy metal ions in wastewater by zeolite 4A and residual products from recycled coal fly ash. *J. Hazard. Mater.*, 127, 1-3, 89–101, 2005.
88. Yuna, Z., Review of the natural, modified, and synthetic zeolites for heavy metals removal from wastewater. *Environ. Eng. Sci.*, 33, 7, 443–454, 2016.
89. Curkovic, L., Cerjan-Stefanovic, S., Filipan, T., Metal ion exchange by natural and modified zeolites. *Water Res.*, 31, 6, 1379–1382, 1997.

90. Ibrahim, H.S., Jamil, T.S., Hegazy, E.Z., Application of zeolite prepared from Egyptian kaolin for the removal of heavy metals: II. Isotherm models. *J. Hazard. Mater.*, 182, 1–3, 842–847, 2010.
91. Panayotova, M., Use of zeolite for cadmium removal from wastewater. *J. Environ. Sci. Health A*, 35, 9, 1591–1601, 2000.
92. Dal Bosco, S.M., Jimenez, R.S., Carvalho, W.A., Removal of toxic metals from wastewater by Brazilian natural scolecite. *J. Colloid Interface Sci.*, 281, 2, 424–431, 2005.
93. Jha, V.K., Matsuda, M., Miyake, M., Sorption properties of the activated carbon-zeolite composite prepared from coal fly ash for Ni2+, Cu2+, Cd2+ and Pb2+. *J. Hazard. Mater.*, 160, 1, 148–153, 2008.
94. Panayotova, M., II, Kinetics and thermodynamics of copper ions removal from wastewater by use of zeolite. *Waste Manag.*, 21, 7, 671–676, 2001.
95. Erdem, E., Karapinar, N., Donat, R., The removal of heavy metal cations by natural zeolites. *J. Colloid Interface Sci.*, 280, 2, 309–314, 2004.
96. Bektaş, N. and Kara, S., Removal of lead from aqueous solutions by natural clinoptilolite: Equilibrium and kinetic studies. *Sep. Purif. Technol.*, 39, 3, 189–200, 2004.

3

Sustainable Green Synergistic Emulsion Liquid Membrane Formulation for Metal Removal from Aqueous Waste Solution

Norasikin Othman[1,2*], Norela Jusoh[1,2], Raja Norimie Raja Sulaiman[1,2] and Norul Fatiha Mohamed Noah[1,2]

[1]*School of Chemical and Energy Engineering, Faculty of Engineering, Universiti Teknologi Malaysia, Johor, Malaysia*
[2]*Centre of Lipids Engineering and Applied Research (CLEAR), Ibnu Sina Institute for Scientific and Industrial Research, Universiti Teknologi Malaysia, Johor, Malaysia*

Abstract

The release of hazardous metal ions beyond the standard discharge limit from chemical industries has posed a great concern to our nation. The emulsion liquid membrane (ELM) technique is among the most promising technologies for metal removal. In this work, several metals, such as silver (Ag), chromium (Cr), nickel (Ni), and zinc (Zn), were treated using different liquid membrane (LM) formulations. Synergism of carriers was employed to enhance the extraction performance while minimizing chemical (carrier) consumption. Results showed that 98% of Ag was synergistically extracted by 0.0002 M Cyanex 302 with 0.0003 M Cyanex 272 in palm oil, with a synergistic factor (SC) of 62.70. Almost 99% of Cr (VI) extraction was achieved using 0.04 M Cyanex 302 + 0.01 M Cyanex 272 with an SC of 7.73. Meanwhile, 83% of Ni extraction was attained by a carrier combination of 0.08 M LIX 63 + 0.02 M D2EHPA with an SC value of 29.56. On the other hand, 80% of Zn was extracted using 1 mM Cyanex 302 + 9 mM D2EHPA with a SC of 2.90. It can be inferred that the developed formulations for synergized ELM extraction could be a sustainable separation method for various metal removals in the future.

Keywords: Emulsion liquid membrane, metal extraction, green formulation, synergistic extraction

**Corresponding author*: norasikin@cheme.utm.my

Shahid-ul-Islam, Abid Hussain Shalla and Mohammad Shahadat (eds.) Green Chemistry for Sustainable Water Purification, (49–78) © 2023 Scrivener Publishing LLC

3.1 Introduction

Rapid industrial activities have caused water resources to be contaminated with heavy and precious metals, dyes, as well as nutrient ions (nitrogen and phosphate) [1]. For instance, the electroplating effluent in the surface water streams is one of the sources of metals. During the electroplating process, different types of metals are used to coat metal objects. A large amount of wastewater containing various base and precious metals is created as a result of the plating and washing process. Treating these polluted sites and recycling the metals has been widely studied to reduce the environmental problem. The commonly used metals in the electroplating industry include silver, chromium, nickel, zinc, copper, tin, gold, and palladium.

Precipitation reaction, coagulation-flocculation, ion exchange, membrane separation, adsorption, electrodialysis, and photocatalytic processes were all considered for wastewater treatment and metal recovery [2, 3]. Unfortunately, these processes suffer from various drawbacks, including sludge generation, high operational costs, and complicated condition requirements. Among the most auspicious methods to remove and recover metals from the electroplating effluent, is the emulsion liquid membrane (ELM) technique. ELM fulfills the assurance of catering to many benefits, such as high efficiency, relatively low power consumption, rapid extraction, the ability to remove numerous substances from a dilute solution, and high selectivity [4, 5].

The right LM formulation composed of diluent, carrier, surfactant, and stripping agent is very crucial to determining the efficiency performance of the targeted solute in the LM technology. Among them, the carrier plays an important function in facilitating the mass transportation of substances in the ELM system. The use of a single carrier is less effective owing to its low loading capacity and slow kinetics of extraction [6, 7]. To enhance the separation performance, a synergistic extraction process could be employed. The process comprises the combination of two or more carriers, wherein the mixed carrier is capable of producing higher performance in comparison to the sum of the individual carriers. The synergism could be accomplished using a combination of different carrier types, for instance, two acidic, two neutral, two basic, anionic-neutral, cationic-ionic, and so on. This could enhance the distribution of the targeted substances in the organic phase. Besides that, the fundamental benefit of mixed carrier systems is that they only utilize the available chemical reagents, thus avoiding the time-consuming development of new reagents.

The efficacy of the synergistic extraction system was demonstrated by Belkhouche *et al.* [8], who obtained almost 100% nickel extraction using a combination of TBP and D2EHPA. Furthermore, the enhancement of solute extraction has also been proven using carrier mixtures such as HDEHP and LIX 860 for the removal of Ni (II) [9], CMPO and PC-88A for the extraction of uranium [10], and LIX 984 with Cyanex 301 for copper extraction [11].

Meanwhile, the ELM approach to sustainable chemistry is possible with the replacement of commercial petroleum-based organic diluents with environmentally friendly vegetable oils [12]. The incorporation of vegetable oils as the diluent in the ELM formulation helps promote sustainable industrial processes as well as minimize the discharge of hazardous materials into the environment. Previously, some researchers have widely studied the application of environmentally friendly vegetable oils in quite a few configurations of liquid membranes, including bulk and supported liquid membranes. For example, Muthuraman and Palanivelu [13] have reported the transport of textile dye utilizing a supported liquid membrane containing a vegetable oil-based formulation. Similarly, Mahmoud *et al.* [14] removed dye from textile wastewater utilizing plant oil in the formulation. Chakrabarty *et al.* [15] have employed coconut oil as the solvent in a liquid membrane formulation for mercury removal. In addition, the incorporation of vegetable oil as the organic phase in the emulsion liquid membrane configuration for metal separation was also reported [16].

This study presents the development of a synergistic LM formulation for the removal of metals such as silver, chromium, nickel, and zinc. The selection of single carrier, mixed carrier, and diluent toward the performance of metal extraction was investigated. Additionally, future challenges and prospects of the ELM process were also addressed.

3.2 Theoretical

Emulsions are stable colloids formed by the dispersion of one liquid into another liquid due to the existence of a surfactant. The emulsion liquid membrane (ELM) was invented from the liquid-liquid extraction process, in which a solute is extracted and stripped simultaneously in one unit operation. In this system, two liquid phases are separated, wherein one phase is confined by a liquid membrane. Principally, the primary water-in-oil (W/O) emulsion consists of an aqueous internal phase (W) and an organic liquid membrane (O) and is prepared. Then, it is dispersed in the aqueous external phase (W), forming a double water-in-oil-in-water (W/O/W) emulsion system. The transportation of the targeted solute by a carrier from the external phase occurs by a reaction at the external interface. Meanwhile, the process of solute stripping takes place at the internal interface.

ELM process flow diagram is presented in Figure 3.1. The ELM operation involves four important operations consist of primary emulsion preparation, dispersion and reaction process, emulsion settling, and demulsification. During the early stage, the stripping agent is emulsified with the organic liquid membrane assisted with a surfactant to obtain primary W/O emulsion. Next, the emulsion is distributed in the external feed solution comprising the targeted substance in a mixer and the extraction process occurs assisted with agitation. During this stage, the targeted solute in the external aqueous solution is transported through the liquid membrane towards the internal stripping phase. The reaction of solute and carrier occurs at the external barrier between the emulsion and the aqueous feed phase. After that, two distinct phases of emulsion and external are being settled by the force of gravity. Finally, the emulsion is emulsified with the purpose to recover the solute and recycle the liquid membrane.

The fundamental principle of ELM separation lies in the reaction of solute and carrier at the external barrier. This complex permeates across the liquid membrane from the external to internal interfaces. Decomposition of this complex occurred at the internal-membrane interface through the exchange of equilibrium extraction and stripping reactions. Then, the

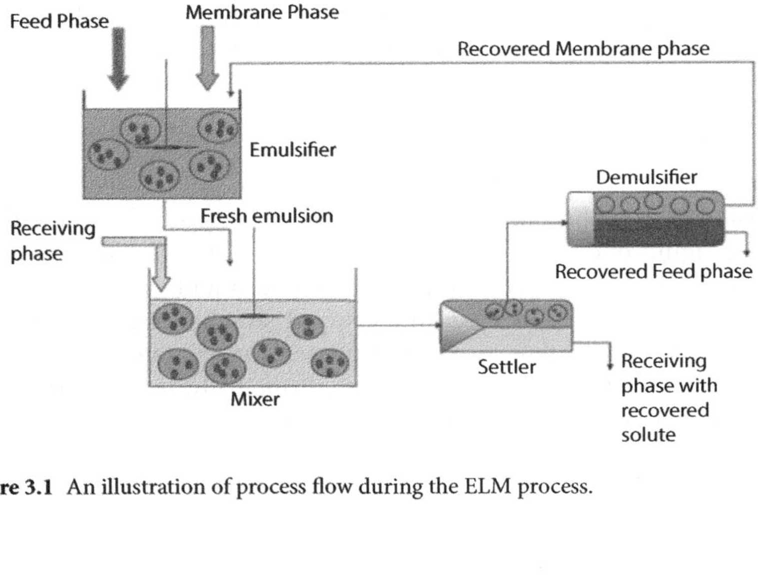

Figure 3.1 An illustration of process flow during the ELM process.

solute ion is confined in the internal stripping phase, and the detached carrier travels again to the external-membrane barrier and reacts with another solute. Therefore, even at a low concentration of solute, ELM can still work appropriately [17, 18].

3.2.1 Mass Transfer Mechanism in the ELM Process

A distinctive liquid membrane is made of an organic solution containing a carrier molecule, diluent, and surfactant. This membrane forms a thin layer between external and internal aqueous phases containing different solute concentrations. The carriers facilitate the transport of solutes across the liquid membrane by altering the permeability of the solutes, thus producing selective transport and separation. Any solute from the aqueous feed phase is transported across the liquid membrane into the internal stripping phase either by diffusion (single transport or passive diffusion) or by simultaneous diffusion and chemical reaction in the liquid membrane phase (coupled transport or facilitated transport) and/or reaction in the internal stripping phase (passive diffusion with stripping reaction) [19].

The ELM systems have been classified into two categories, specifically type I and type II, depending on the existence of a carrier within the liquid membrane phase. The chemical potential difference between the external feed and internal stripping phases drives the transportation of the targeted substance through the liquid membrane [20]. Besides, the solutes are transported from a higher to a lower concentration area.

In the type I ELM system, the targeted substance can diffuse in all liquid phases, including external liquid membranes, as well as internal phases. Hence, passive diffusion of the targeted substance toward the internal stripping phase across the liquid membrane can be achieved. The reaction between the stripping agent and the solute normally arises at the internal barrier amid the stripping phase and liquid membrane, producing an insoluble substance. As a consequence of this stripping process, the stripped product is unable to travel back into the liquid membrane phase. Consequently, the reaction also keeps the lowest concentration of solute in the internal phase.

For instance, benzoic acid separation from an aqueous external feed phase is applied to a type I ELM system [21]. The benzoic acid initially dissolves in the external aqueous feed phase. Then, it is transported by diffusion across the organic liquid membrane and stripped by sodium hydroxide from the internal phase. In the internal phase, the benzoic acid exists in the form of a benzoate ion that is insoluble in the organic liquid membrane. Thus, the benzoate ions are confined to the internal stripping phase. In the meantime, the benzoic acid concentration remained zero in the stripping phase, hence providing a high chemical potential difference between benzoic acid in the external and internal phases. The benzoic acid continues to be transferred until it is completely extracted. Meanwhile, the sodium hydroxide molarity determines the capacity of the internal phase.

In the type II ELM system, the targeted solute is unable to dissolve in the organic liquid membrane phase and resists diffusional mass transfer. As a result, another component known as a "carrier" is added to the organic liquid membrane to aid mass transportation. During the extraction process, the carrier and solute react at the external barrier between the liquid membrane and the external phase. The solute is confined in the internal phase by a stripping process at the internal barrier, releasing free carriers that diffuse to the external interface and react with another solute.

An example of the type II ELM mechanism is the extraction of zinc (Zn) from the aqueous external feed phase [21]. First, zinc in the aqueous external feed solution reacts with the carrier agent presented in the liquid membrane at the external barrier and forms zinc-carrier complexes. The complexes formed are soluble in the liquid membrane and transported to the internal phase. At the internal barrier, zinc is detached from the complexes and confined to the internal phase by a stripping reaction. The reaction converts the complex into two compounds, one being the carrier itself and the other being zinc that is insoluble in the liquid membrane phase. The carrier travels again into the liquid membrane phase and participates in the subsequent reaction for complex formation, thus speeding up the removal rate of zinc.

The concentration gradient of the solute-carrier complex between the external and internal interface drives the mass transfer in the type II ELM

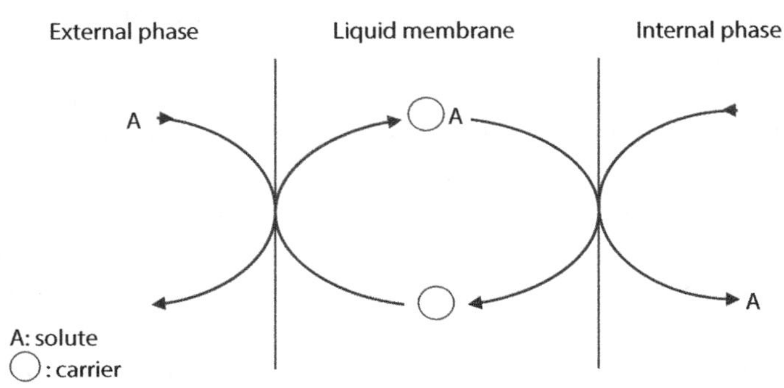

Figure 3.2 The coupled transport mechanism in the type II ELM system.

mechanism. Apart from the complex concentration, the pH difference also drives solute transportation [22]. According to its advantage of transporting a feed solute which is insoluble in the membrane phase, type II ELMs were applied in this study. The detailed scheme for the diffusion of substances from the external phase across the liquid membrane phase and further accumulation in the internal phase is illustrated in Figure 3.2.

3.2.2 Component Selection in the ELM

The ELM process could be used to fulfill the requirement of different extraction processes through potential liquid membrane formulation for various ions selectivity from the external solution. The development of the ELM process is a straightforward procedure, but maintaining the system for long-term use is complicated. To obtain an effective separation, the emulsion must possess neutral buoyancy and suspends in the external phase, have high viscosity, and be stable [23]. Hence, the liquid membrane formulation is important which comprises the screening of carriers, diluents, surfactants, and strippants. The selection of components in the ELM formulation regulates the performance of the ELM process. Generally, the carrier must have a high selectivity to the targeted substance in the external aqueous phase, while the appropriate selection of stripping agent and surfactant must be considered to avoid the cotransport of the aqueous phase and unwanted compounds during the ELM extraction.

Throughout the ELM process, a specific carrier is required to react with the solute that is selected according to certain criteria. The formulation is tailor-made and various carriers were screened to find the most appropriate LM component. For instance, any substance, molecule, or atom that can react with solute could serve as a carrier, as long as it is immiscible with the second liquid phase (i.e., aqueous phase). The interaction of carrier with solute generally involves solvation, ion association, and compound formation. There are three categories of carriers, namely acidic, basic, and solvating. The acidic carriers are organic substances that are normally derived from monocarboxylic acids and phosphorous acids. Besides, the esters of phosphinic, phosphonic, and orthophosphoric acids were included. For example, the alkyl phosphoric acid group is D2EHPA, while Cyanex 272 is from the phosphinic acid group. These carriers react with cationic solute via a cation exchange reaction whereby the hydrogen of the carrier is replaced by the solute ion.

Basic carriers are substances containing amines and quaternary ammonium halides groups. The amines are categorized as primary, secondary, tertiary, and quaternary ammonium salts. Some examples of these carriers are tri-octylamine, tri-dodecyl amine, and Aliquat 336, which have shown a tendency to extract organic molecules such as reactive dyes and organic acid [24, 25]. Meanwhile, solvating carrier involves solvation and mainly depends on the role of carrier-containing oxygen molecules to solubilize inorganic complexes or molecules. There are two main groups of solvating carriers which are oxygen or sulfur bonded to phosphorus and organic reagent consisting of oxygen-carbon bonds. Generally, the selection of carrier should have a thermodynamic preference towards solute at the interface. Besides that, the reaction should be kinetically fast for high extraction efficiency. In this study, a few types of carriers and their combination were screened for the extraction of metals ion.

In terms of diluent, the characteristic features of diluent are polarity, viscosity, water miscibility, volatility, and surface tension. The good features of the diluents should have high viscosity and should be cheap. Stripping agent is another component in LM formulation. The choice of the stripping agent is predominantly influenced by the carrier used. Normally, extraction of metal cation which employs acidic carriers requires an acidic stripping agent [26]. Acidic stripping agent seems to be suitable for acidic carrier used since they are capable of dissociating the solute-acidic carrier complexes. Meanwhile, the extraction of anionic complexes, which use basic or solvating carriers needs basic or neutral stripping agents [27]. Generally, the basic carrier type contains an amine group that extracts metal ion complexes via an anion exchange mechanism. To exchange the anion with the solute, the molecular amine carrier must be first transformed to the amine salt.

Many studies have been made on ELM extraction of metals, for instance, arsenic, bismuth, cadmium, chromium, cobalt, silver, nickel, palladium, and tungsten. Kiani and Mousavi [28] had successfully removed arsenic from water using Cyanex as a carrier, paraffin as a diluent, sodium sulfate as a strippant, and Span 80 and Tween 80 as surfactants. Benyahia *et al.* [29] and Mokhtari and Pourabdollah [30] optimized extraction of bismuth, Bi(III) using di(2-Ethylhexyl) phosphoric acid (D2EHPA) as a carrier, sulphuric acid, H_2SO_4 as a stripping agent, and iso-octyl phenoxy poly ethoxy ethanol (Triton X-100) as a surfactant. Their study presented that almost 100% of Bi(III) was extracted at the optimal condition. ELM removal of cadmium and chromium mostly used a basic carrier from amine group, including trioctylamine (TOA), tri-n-octyl ammonium chloride (TOMAC) (also called Aliquat 336) [31, 32]. Petroleum-based kerosene is mainly employed as a diluent for chromium extraction aside from the use of vegetable oil which becomes of interest recently. Meanwhile, sodium hydroxide (NaOH) and sorbitan monooleate (Span 80) were usually used in chromium extraction as stripping agents and surfactants, respectively.

Kumbasar [33] successfully extracted cobalt, Co(II) from acidic and basic solutions employing tri-iso-octyl amine (TIOA) and 8-hydroxyquinoline (8-HQ), respectively. Besides, for both cases, kerosene was utilized as a diluent, H_2SO_4 as a strippant, and nonionic polyamine (ECA 4360J) as a surfactant. Sulaiman *et al.* [34] demonstrated 100% of nanosilver was extracted from wash water adopting bis [2,4,4-trimethyl pentyl] monothio-phosphinic acid (Cyanex 302) as a carrier, Span 80 as a surface-active agent, kerosene as a solvent, and acidic thiourea as a strippant. Eyupoglu and Kumbasar [35] extracted nickel, Ni(II) from spent chromium-nickel electroplating effluent using aliphatic 5,8-diethyl-7-hydroxy dodecane-6-one oxime (LIX 63) and 2-bromo decanoic acid (2BDA) as carriers ELM

Table 3.1 LM formulation for metal extraction.

Metal	Carrier	Diluent	Stripping agent	Surfactant	References
Arsenic (V)	Cyanex 921	Paraffin	Na_2SO_4	Span 80 Tween 20	[28]
Bismuth, Bi(III)	D2EHPA	n-pentanol	H_2SO_4	Triton X-100	[30]
Bismuth, Bi(III)	D2EHPA	Dichloro-methane	H_2SO_4	Triton X-100	[29]
Chromium	TOMAC/Aliquat 336	Kerosene-palm oil	NaOH	Span 80	[37]
Chromium, Cr(IV)	TOMAC/Aliquat 336	Kerosene	NaOH	Span 80	[31]
Cobalt, Co(II)	Triisooctylamine (TIOA)	Kerosene	H_2SO_4	Span 80	[33]
Nanosilver	Cyanex 302	Kerosene	Thiourea in H_2SO_4	Span 80	[34]
Palladium	Cyanex 302	Kerosene	Thiourea H_2SO_4	Span 80	[17]
Tungsten	TOMAC/Aliquat 336	Hexane	NaOH	Span 80	[36]

up to 98.5%. Their study found that 2BDA acted as a synergist in the process. Noah [17] selectively transported almost 100% of palladium, Pd from electroplating bath solution using Cyanex 302 as a carrier, kerosene as a solvent, acidic thiourea as a strippant, and Span 80 as a surfactant. Meanwhile, Lende and Kulkarni [36] recovered tungsten or wolfram from the rinsing wastewater of printed circuit board. The maximum tungsten recovered from this study was 80% employing Aliquat 336 as a carrier in hexane, NaOH as a strippant, and Span 80 as a surface-active agent.

Some of the ELM formulations for metal extraction as reported in a previous study are summarized in Table 3.1. The formulation is regarded as tailor-made due to the selectivity to the targeted solute according to certain criteria.

3.3 Experimental

3.3.1 Materials

Autocatalytic nickel plating effluent was obtained from Seagate International Sdn Bhd based in Johor, Malaysia. Cr(VI) plating wastewater was supplied by Perstima Sdn Bhd in Pasir Gudang, Johor, Malaysia. Nitric acid (HNO3, purity 65%), kerosene, di-2,4,4-trimethyl pentyl mono-thiophosphinic acid (Cyanex 302) (99% assay), di-2,4,4,-trimethyl pentyl phosphinic acid (Cyanex 272) (90% assay), Di-2-Ethylhexyl phosphoric acid (D2EHPA) (≥95% assay) were ordered from Sigma Aldrich. 5,8-diethyl-7-hydroxy dodecane-6-one oxime (LIX63, purity 99%) was supplied by Cognis, Australia. Refined palm cooking oil was obtained from Lam Soon Edible Oils (Brand: Buruh). Silver nitrate solution ($AgNO_3$) (99% assay), Zinc chloride ($ZnCl_2$) (99% assay), hydrochloric acid (HCl) (37% assay), and thiourea (99% assay) as strippants were bought from Merck. All the reagents and chemicals were directly used as obtained as it is disregarding any further refinement.

3.3.2 Reactive Extraction Procedure

The LM components selection for metal removal was identified using a reactive extraction method. An aqueous feed phase (metal solution) and organic phase (carrier in diluent) were combined in equal proportions in a conical flask and mixed using a rotary flask shaker at 300 rpm agitation speed and 18 hours to attain extraction equilibrium. Once the mixing procedure was completed, the solution was transmitted cautiously into a separatory funnel and allowed to settle at room temperature (25°C) for 30 minutes. The aqueous phase formed at the bottom of the separatory funnel was collected for metal concentration determination using atomic absorption spectrometry (AAS). The efficiency of the extraction process was evaluated based on the mass balance technique.

To evaluate the synergistic effect, an equivalent amount of aqueous feed phase was mixed with organic phase (mixed carrier in diluent) and shook in the rotary flask shaker for 1 hour at 300 rpm. Investigations on the effect of individual and mixed carriers, together with their concentrations were carried out employing a similar method. The aqueous and organic phase settling, as well as the determination of metal concentration, were performed utilizing the aforementioned procedure. The overall metal extraction procedure using single and mixed carriers is shown in Figure 3.3.

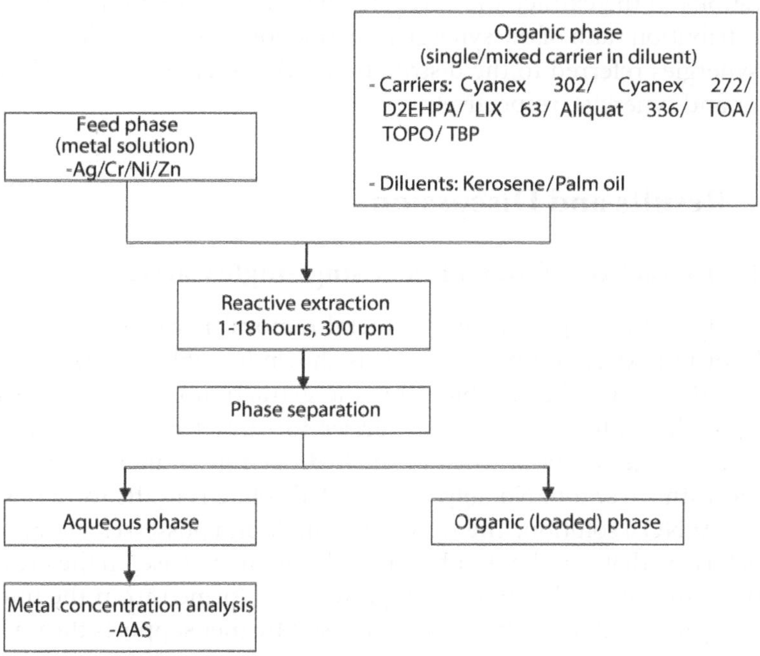

Figure 3.3 Overall metal extraction procedure using single and mixed carriers.

3.3.3 Determination and Calculations

The metal extraction performance, distribution ratio (D), and synergistic factor (SC) were determined employing Equations 3.1–3.3, accordingly.

$$\text{Extraction}(\%) = \frac{C_i - C_f}{C_i} \times 100 \tag{3.1}$$

$$\text{Distribution ratio, } D = \frac{C_{org}}{C_{aq}} \times 100 \tag{3.2}$$

$$\text{Synergistic factor, } SC = \frac{D_{mixture}}{D_{carrier} + D_{synergist}} \tag{3.3}$$

where C_i and C_f are the initial and final metal concentrations (ppm) in the feed phase for the extraction process. C_{org} and C_{aq} denote as the metal concentrations in the extract and raffinate phases, correspondingly. $D_{mixture}$ is the distribution ratio of the synergistic extraction process, while subcarrier and synergies referred to the distribution ratio obtained using individual carrier and synergist, respectively.

3.4 Results and Discussion

3.4.1 Extraction of Metal Ions Using Single Carrier

The study of the single carrier for metal removal is carried out and the result for the extraction performance is shown in Table 3.2. As observed, Cyanex 302 shows the capability for the extraction of Ag. The result is owing to the ability of the acidic carrier to extract metals ion through cation-exchange reaction, wherein the hydrogen in Cyanex 302 was substituted with silver ion. Notably, the targeted solute is in the form of silver nitrate ($AgNO_3$) solution. The $AgNO_3$ is soluble and possesses a lipophobic characteristic that could not diffuse into the organic phase. In the presence of carrier molecules, lipophilic complexes were formed from the interaction of Cyanex 302 with silver ions. The result further supports the findings of the previous studies that reported the viability of employing Cyanex 302 in the extraction of metal [38, 39].

Table 3.2 LM formulation using a single carrier (Experimental conditions: stirring speed: 300 rpm; extraction time: 18 hours).

Metals	Carrier	Diluent	Performance (%)
Ag	0.2 mM Cyanex 302	Palm oil	45.4
Cr	0.10 M Cyanex 302	Kerosene	68.0
Ni	0.10 M LIX63	Kerosene	13.0
Zn	0.01 M D2EHPA	Kerosene	64.0

On the other hand, Cyanex 302 also shows the ability to extract Cr(VI) from acidic waste solution. Such a condition was strongly supported by literature that claimed that Cyanex 302 is a strong acidic carrier that can extract the metal from acidic aqueous solution [40, 41]. Sole and Hiskey [40] also revealed that the acidity of Cyanex 302 increases due to the low acid dissociation constant (pKa) where pKa of Cyanex 302 (5.63) > Cyanex 272 (6.37). Besides, Cyanex 302 also could reduce Cr(VI) to Cr(III) [41]. In terms of reaction, the shared electron between metal-sulfur bonds is more stable compared to metal-oxygen, thus enhancing the strength of the bonding [42].

Meanwhile, Chauhan and Patel [43] revealed that LIX63 offers low extraction kinetics but excellently extracts nickel from acidic waste solution. Normally, LIX63 is commonly used with the mixture of other carriers, hence raising a synergism effect toward nickel extraction from acidic feed solution [44].

As for zinc extraction, the high affinity of D2EHPA might be explained by the existence of oxygen-phosphorous chemical bonding and alkyl radicals which results in its strong acidic character (phosphoric acid) [45]. In addition, pH 4.9 of zinc solution provides a lower proton concentration, hence resulting in high extraction efficiency. This phenomenon generally occurs during the extraction of cationic metals using the acidic carrier including D2EHPA.

3.4.2 Extraction of Metal Ions Using Mixed of Carriers

The results of metal ions extraction from a single carrier show the suitability of the employed carrier accordingly. However, the performance of extraction was unfavorable, which is only around 13% to 68% of extraction was achieved. One way to enhance the extractive capacities of the carrier and minimize the amount of carrier used is through the employment of a synergistic extraction system. The synergistic effect can be achieved through a combination of two carrier types that are capable of enhancing extraction efficiency. The result for the synergistic mixture of carriers in LM for metal removal is shown in Table 3.3.

The result shows that the combination of Cyanex 272 with Cyanex 302 enhanced the extraction of silver ions. Principally, an anionic substance is needed to generate complexes with the metal ions in the aqueous feed. Consequently, a combination of acidic carriers is very effective to extract silver ions through the cation exchange reaction. It can be seen that more than 98% of silver ions extraction was achieved using 0.2 mM Cyanex 302 and 0.3 mM Cyanex 272. Besides, a high synergistic coefficient value was attained which is 62.70.

According to the preliminary reported results, Ag reacts with Cyanex 302 and Cyanex 272 at the ratio of 10:20:1 [46]. This is based on the stoichiometric plot of log Dmixture versus log carrier concentration. As can be seen from Figure 3.4, the slope is around 0.12, implying the quantity of Cyanex 272 participated in the reaction is very small. The result of this study suggests that Cyanex 272 functions as a transporter to carry complexes of silver ions in the organic phase. Referring to Figure 3.5, the slope value of 4.1 was obtained, indicating that 4 moles of Cyanex 302 were required to react with silver ions. According to the general principle of

Table 3.3 LM formulation using a mixed carrier (Experimental conditions: extraction speed: 300 rpm; extraction time: 1 hour).

Metals	Carriers	Diluent	Performance (%)	Synergistic factor (SC)
Ag	0.0002 M Cyanex 302 + 0.0003 M Cyanex 272	Palm oil	98	62.70
Cr	0.01 M Cyanex 272 + 0.04 M Cyanex 302	Kerosene	99	7.73
Ni	0.08 M LIX63 + 0.02 M D2EHPA	Palm oil	83	29.56
Zn	1 mM Cyanex 302 + 9 mM D2EHPA	Kerosene	80	2.90

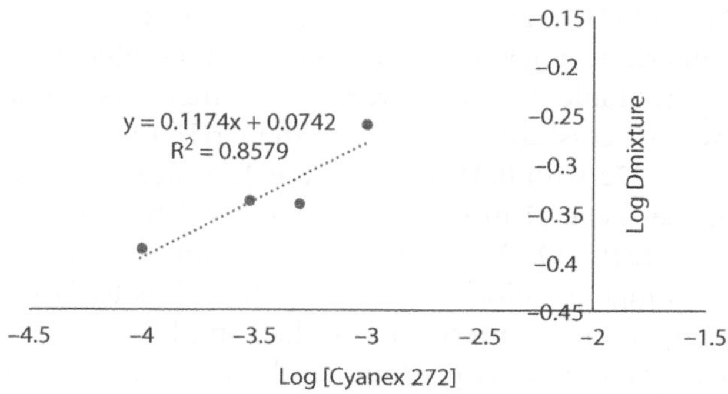

Figure 3.4 Stoichiometric study of mixture system on silver ions extraction at a rigid Cyanex 302 concentration (diluent: palm oil; extraction speed: 300 rpm; extraction time: 1 hour; temperature: 25°C).

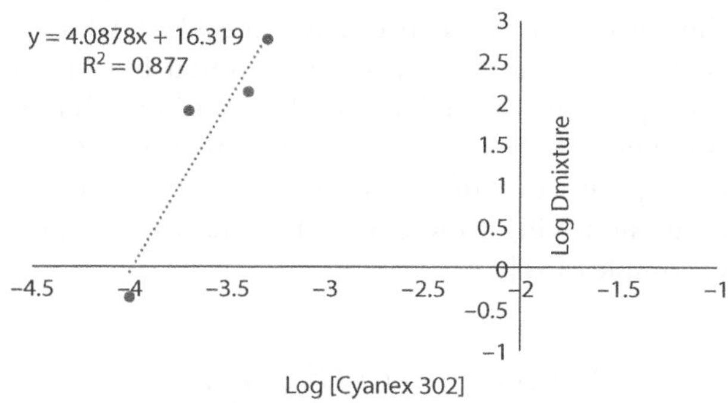

Figure 3.5 Stoichiometric study of mixture system toward the extraction of sliver at a rigid Cyanex 272 concentration (diluent, palm oil; extraction speed, 300 rp; extraction time, 1 hour; temperature, 25°C).

kinetic, the foremost carrier usually gives low extraction kinetics, while the synergist carrier provides rapid extraction and kinetics. Therefore, this finding suggests that Cyanex 302 and Cyanex 272 are designated as a carrier and synergists, respectively.

The proposed extraction mechanism of Ag ions by the synergistic carrier is shown in Equation 3.4. Besides, the investigation of the individual and carriers mixture indicates Cyanex 302 and Cyanex 272 serve as a carrier and synergist, respectively.

$$10Ag^+ + 20(HA)_2 + (HB)_2 \rightarrow 10(AgHA_2) + 10(HA)_2 \cdot (HB)_2 + 10H^+ \tag{3.4}$$

where $(HA)_2$ and $(HB)_2$ are Cyanex 302 and Cyanex 272 that commonly exists in a dimeric arrangement in the low polarities diluent [47].

According to Table 3.3, compared to the single system, almost 99% of Cr(VI) was successfully extracted to the Cr(III) via combinations of 0.01 M Cyanex 272 with 0.04 M Cyanex 302. It means this mixture system pronounces synergism effect with a synergistic coefficient (SC) of 7.73. Dichromate ($Cr_2O_7^{2-}$) is dominant for the aqueous phase containing Cr which is orange in color [48]. $HCrO_4^-$ dimerizes to form $Cr_2O_7^{2-}$, as denoted by Equation 5. In terms of mechanism, Liu *et al.* [49] reported that the synergism effect is predominantly influenced by the property of the solute and the organic phase. Cyanex 302 offers a double role in this system. First, Cyanex 302 tends to reduce the pK_a value of Cyanex 272 in the organic phase for the maximum Cr(VI) separation from an acidic aqueous feed phase [50]. Second, Cyanex 302 acts as a reducer of Cr(VI) to Cr(III) [40, 42]. As for reaction, hydrogen chromate ion (HCrO4-) dimerizes to form dichromate (Cr2O72-), which is dominant for the aqueous phase containing Cr as denoted by Equation 3.5. An equation that represents the reduction of chromium after reacting with ($R_2P(S)OH$) (Cynex 302) is shown in Equation 3.6, Figure 3.6(a) illustrates the molar proportion of Cyanex 302: Cr is 2:1 in the solute-carrier complexes, thus signifying 1 mole of Cr interact with 2 moles of Cyanex 302 forming complex [51].

$$2HCrO_{4(aq)}^{-} \rightarrow Cr_2O_{7(aq)}^{2-} + H_2O_{(aq)} \tag{3.5}$$

$$\begin{aligned} &Cr_2O_{7(aq)}^{2-} + H_2O_{(aq)} + 2\big(R_2P(S)OH)_{2(org)} \\ &\quad \rightarrow 2(R_2P(OH)S^-)_2.Cr(OH)_2^+\big)_{(org)} + 3O_{2(g)} \end{aligned} \tag{3.6}$$

Conferring to the hard-soft-acid-base (HSAB) principles, soft-soft and hard-hard interactions form the most stable molecule complexes [49]. Thus, Cyanex 272 (hard base) forms a stable complex with reduced Cr (III) ions (hard acid) [52]. Figure 3.6(b) shows the stoichiometric of Cyanex 272 by constructing a graph of log D_{mix} versus log [Cyanex 272]. The slope of 0.2 suggests that the molar proportions of Cyanex 272: Cr is 1:5 in the solute-carrier complexes as represented by Equation 3.7 [51]:

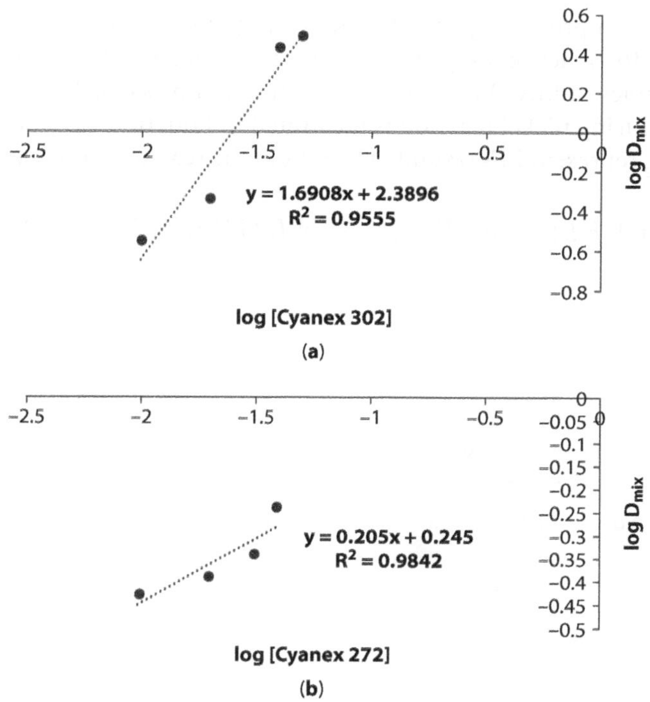

Figure 3.6 Stoichiometric study of mixture system at fixed (a) Cyanex 272 and (b) Cyanex 302 concentrations towards chromium extraction. Experimental conditions: diluent: kerosene; [Cr]: 41 ppm; agitation time: 18 hours; agitator speed: 320 rpm; temperature = 26 ± 1 °C.

$$2\left(R_2P(OH)S^-)_2.Cr(OH)_2^+\right)_{(org)} + 5(R_2P(O)(OH)_{2(org)}$$
$$\rightarrow 2(R_2P(O)O^-)_2.Cr(OH)^+_{2(org)}$$
$$+ 2R_2P(S)OH\ldots(^-OOPR_2)_{3(org)} + 10H^+ \quad (3.7)$$

Thereafter, a synergistic system involving a mixture of D2EHPA and LIX63 has been studied for nickel removal from electroplating wastewater. Through the synergistic process, 83% of nickel was successfully extracted using 0.08 M LIX63 with 0.02 M D2EHPA with a high SC value of 30. In terms of synergism, the mixture of LIX63 and D2EHPA can transport a higher number of hydrogen atoms to the aqueous feed phase which is later replaced by the nickel ions [53]. Then, the stoichiometric reaction was studied through a graph of log Dmix versus log [D2EHPA]. The slope of the graph is approaching 1.0, thus indicating that 1.0 mole of D2EHPA is needed in the reaction as shown in Figure 3.7. Meanwhile, Figure 3.8 proposes 4 moles of LIX63 needed in the extraction process. The mechanism involved during nickel reaction is illustrated in Equation 3.8 wherein (HL) and $(HR)_2$ represent LIX 63 and D2EHPA in the reaction, respectively [54].

$$Ni^{2+}_{(aq)} + 4(HL)_{(org)} + (HR)_{2(org)} \rightarrow NiL_2(HL)_2(HR)_{2(org)} + 2H^+ \quad (3.8)$$

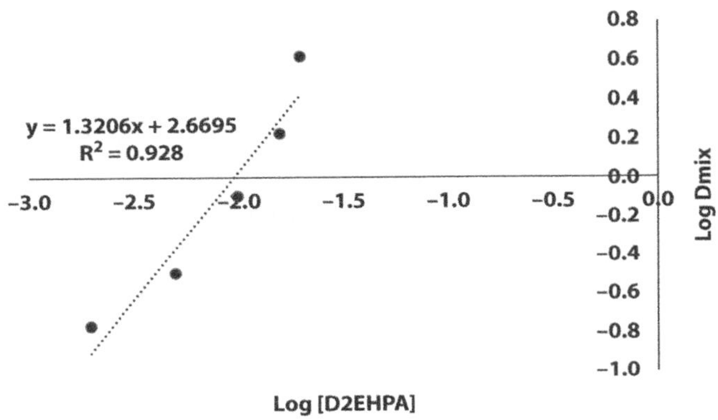

Figure 3.7 Stoichiometric of D2EHPA towards nickel extraction at a rigid LIX63 concentration. Experimental conditions: Nickel concentration: 500 ppm, pH: 4.84; LIX63 concentration: 0.1 M.

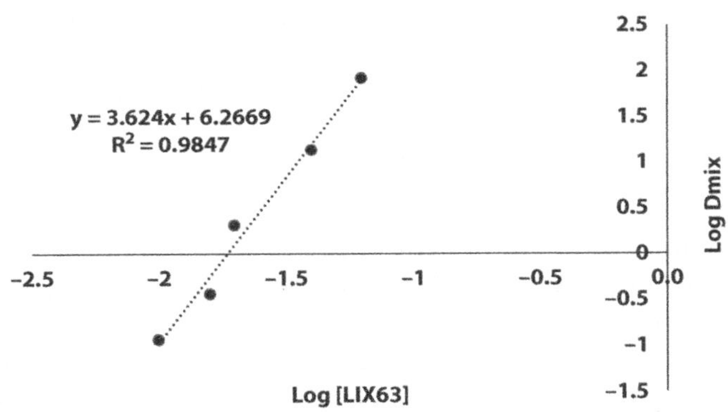

Figure 3.8 Stoichiometric of LIX63 towards nickel extraction at a rigid D2EHPA concentration. Experimental conditions: Nickel concentration: 500 ppm, pH: 4.84; D2EHPA concentration: 0.1M.

Meanwhile, the combination of D2EHPA/Cyanex 302 increased the zinc extraction performance. Results show that at a concentration of 1 mM Cyanex 302 + 9 mM D2EHPA, 72% of zinc was extracted with a synergistic coefficient (SC) of 2.9. This is due to Cyanex 302 being a mono-thio configuration of Cyanex 272 possessing low metal extraction level and slow phase disengagement properties.

The plot of log D against log carrier was constructed to find synergistic reactive extraction equilibrium for zinc extraction. Based on Figure 3.9, the slope of the graph log D_{mix} vs. log $[HA]_{org}$ is 0.98 implying that 1 mole of D2EHPA is required extraction with 1 mole of Zn (II). The number of moles of Cyanex 302 required for the reaction is determined using a similar method. A graph of log D_{mix} versus log $[HB]_{org}$ is constructed, while maintaining the other parameters at constant. Conferring to Figure 3.10, the slope of the graph is 1.18, indicating that 1 mole of Cyanex 302 involves in the extraction of zinc.

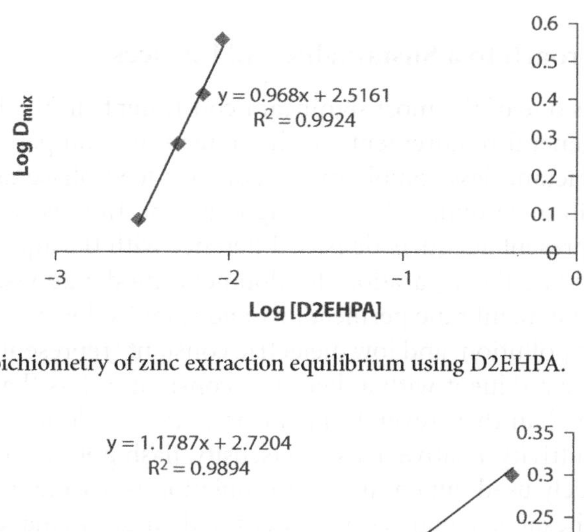

Figure 3.9 Stoichiometry of zinc extraction equilibrium using D2EHPA.

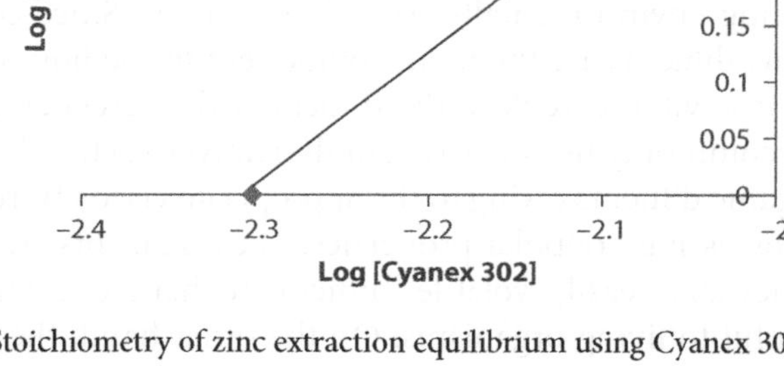

Figure 3.10 Stoichiometry of zinc extraction equilibrium using Cyanex 302.

When Cyanex 302 is added to D2EHPA as a synergist carrier, it helps to improve the extraction performance of zinc by disassociating dimeric D2EHPA to two monomeric D2EHPA molecules (Equation 3.9). Hence, the capacity of D2EHPA increases due to its disassociation that overcomes the less extraction capacity of dimeric D2EHPA. In the D2EHPA and Cyanex 302 mixture, the influence of D2EHPA as an extractor is more prominent compared with Cyanex 302. From this reactive extraction process, Cyanex 302 functions as a synergist. The reaction equation of zinc with the D2EHPA/Cyanex 302 can be expressed in Equation 3.10:

$$2HA + 2HB \rightarrow 2H^{+} + 2\,[A^{-}\,.\,HB] \tag{3.9}$$

$$Zn^{2+}_{(aq)} + 2[A^{-}.HB]_{(org)} \rightarrow Zn.A_2.2HB_{(org)} + 2H^{+}_{(aq)} \tag{3.10}$$

where $2HA$ and $2HB$ are the dimeric form of D2EHPA and Cyanex 302.

3.4.3 Approach to a Sustainable ELM Process

The diluent is one of the most significant constituents in the ELM formulation. The general requirements of the diluent are compatible with carrier and surfactant, less soluble in the aqueous feed phase and receiving phase (to minimize diluent loss during emulsification as well as solvent extraction), present adequate density difference with the aqueous external phase to facilitate the separation development, moderate viscosity which can balance the membrane permeability and stability, low price, low toxicity to prevent pollution, and low dielectric constant (represents the diluent polarity where a diluent with a dielectric constant of less than 15 is nonpolar). Other than that, several diluent properties include melting point, relative permittivity, relative density, viscosity, flash point, and vapor pressure. The widely used diluent in LM formulation is petroleum-based such as kerosene, hexane, heptane, dodecane, and dichloromethane.

An emulsion liquid membrane could be adapted to a sustainable system by incorporating environmentally friendly resources. Since conventional volatile organic diluents are the main component in the liquid membrane formulation, it is vital to replace the diluent with a greener component. Most studies commonly used petroleum derivatives such as kerosene and hexane as organic diluents owing to the appropriate viscosity, readily availability, and possessing nonpolar properties. These reagents are recognized to be nonrenewable, easily volatile, difficult to handle, highly flammable, and harmful to living organisms. On the other hand, the ELM's total operating costs could be affected as a result of price discrepancy owing to a restricted source of petroleum-based diluent.

To promote green chemistry, environmentally friendly materials are incorporated in certain LM formulations. In this work, kerosene and vegetable-based palm cooking oil were used as the diluent as stated in Tables 3.2 and 3.3. It can be seen that both materials are suitable to be used in the LM formulation owing to the similar properties possessed. For example, kerosene and palm cooking oil are almost completely immiscible with water that provides good phase separations in the aqueous phase. Other properties comprise high flash point, low dielectric constant, a lower melting point, and good density gap with the aqueous phase, which are the general requirements for a good diluent. Meanwhile, the viscosity of palm oil is higher compared to kerosene. This could be attributed to the intermolecular interaction in the long-chain fatty acid in palm oil. The higher viscosity is usually unfavorable to avoid resistance of solute transport across the LM phase. In this study, it was found the viscosity of palm oil gives an insignificant outcome on the metals extraction performance. Following the present results, previous research has demonstrated that vegetable oil can be used as a diluent [12, 55].

3.4.4 Prospect and Future Challenges in ELM Technology

Synergism among carriers has become a new perspective employed in the LM formulation for various metal extractions. Through synergism, several good impacts can be observed, including the reduction of a single system's total concentration and fast reaction. For instance, the extraction of silver using a combination of 0.0002 M Cyanex 302 + 0.0003 M Cyanex 272 gives 98% extraction, which is greater compared to the sum of the discrete carrier effects of 0.0002 M Cyanex 302 (45.45%) and 0.0003 M Cyanex 272 (19.16%). Throughout synergistic extraction of chromium, a 50% reduction of the individual carrier concentration (0.10 M) was achieved when employing the mixed system (0.05 M) containing 0.01 M Cyanex 272 and 0.04 M Cyanex 302. Cyanex 302 as a synergist helps reduce the Cr (VI) to Cr (III) before being extracted by Cyanex 272 into the organic phase.

Referring to data for nickel extraction, there is no change in terms of the total concentration of single and mixture systems. However, it can be perceived that there is a significant gap in terms of the extraction performance wherein a carrier combination of 0.08 M LIX63 and 0.02 M D2EHPA has provided 83% of nickel extraction compared to the total concentration of individual carriers of 0.10 M LIX63 (13%) and 0.10 M D2EHPA (33%). Additionally, this system equilibrium can be achieved within 20 minutes only. However, there are similarities between the D2EHPA and Cyanex 302 mixtures for zinc extraction, with a mixture of 1 mM Cyanex 302 + 9 mM D2EHPA providing 78% zinc extraction compared to the total concentration of a single carrier system of 0.01 M D2EHPA (52% + 0.01 M Cyanex 302 (64%). The finding indicates that the transport of solute-carrier complexes is enhanced using synergistic reactive extraction.

To extend the application of synergistic LM formulation in the ELM process, several aspects must be considered. For instance, the determination of ELM optimum conditions has proven to be complicated. This is because several factors affecting the ELM process must be considered, including emulsification time, emulsification speed, surfactant concentration, agitation speed, carrier, and strippant concentration, treatment ratio, extraction time, etc. The optimization of these parameters is important to obtain excellent high-removal and recovery performance.

For instance, ELM efficiency is influenced by emulsification time. If insufficient time was provided for the emulsification time, the emulsion would tend to break immediately. At a longer emulsification time, a small size of emulsion will be produced, thus providing a larger area for solute transportation and enhancing extraction performance. Besides, this condition also enhances the solution intensity and improves emulsion stability [17]. This is owing to the increasing homogeneity of the emulsion and the fact that more stripping agents will be confined to the liquid membrane phase. Therefore, the interfacial area contact will be increased owing to the finer droplets produced. However, prolonged emulsification time over a certain limit leads to undesirable extraction performance as a consequence of the increment in emulsion viscosity and emulsion swelling [56].

Besides that, the extraction efficiency in the ELM process is influenced by emulsification speed. At a lower speed, the energy provided to break the emulsion is insufficient to produce a larger droplet with a small interfacial area for mass transfer. Meanwhile, higher emulsification speeds enhance the extraction performance by virtue of the production of smaller emulsion droplets with a larger interfacial area for mass transport [57]. A further increase in emulsification speed might decrease the extraction performance due to the excess energy provided that leads to emulsion rupture.

An additional significant parameter in the ELM system is the surfactant that controls the stability and affects the permeation of solutes in the liquid membrane. A small amount of surfactant introduced during the emulsification process causes difficulty in emulsion dispersion due to high surface tension [58]. According to Kislik [59], the extraction efficiency increased with surfactant concentration wherein smaller droplet sizes were produced as a result of lower interfacial tension. Consequently, a larger mass transfer interfacial area was provided and enhanced the extraction performance. Research by Othman *et al.* [60] asserted the increase of Span 80 (surfactant) concentration in phenolic compound removal resulted in high extraction performance due to high contact and area of mass transport. Meanwhile, the addition of surfactant concentration beyond a certain limit reduced the extraction performance. This is due to the emulsion tending to swell and break, which is attributed to water cotransport from the aqueous feed phase to the emulsion phase [61].

Furthermore, agitation speed is another important factor in the ELM process. Several studies [62] found that the performance of ELM extraction improves with increasing agitation speed. Generally, a higher agitation speed promotes the production of larger emulsion globules, consequently providing a larger interfacial area amid the emulsion and aqueous feed phase. Subsequently, the rate of mass transport is increased, which enhances the extraction performance. Nevertheless, the increase in agitation speed exceeding a certain level causes emulsion rupture and breakage, thereby obstructing the extraction performance and diluting the concentration of the recovered substance in the internal phase. A short review by Jusoh *et al.* [63] indicated that 200 to 300 rpm of agitation speed is adequate for the ELM process.

In addition, the carrier and its concentration also significantly influence ELM process efficiency. Essentially, the exploration of the carrier effect in the ELM process is initiated by the screening of suitable carriers and variation of their concentration. Yet, a study by Sulaiman *et al.* [64] reported that no significant effect on extraction efficiency was observed after increasing the carrier concentration beyond the optimum level. This is because the viscosity of the emulsion was higher and caused a higher mass transfer resistance, thus the extraction performance was affected [65].

Meanwhile, another factor that affects the ELM's performance is stripping agent concentration. Generally, the capacity of the internal phase for solute recovery increases with stripping agent concentration, consequently increasing the extraction performance [59]. In contrast, the ELM performance was reduced at stripping agent concentration beyond a certain limit on account of emulsion instability owing to the swelling of emulsion that occurs due to the osmotic pressure gap amid the internal and external phases [15].

Besides that, the volume fraction of the emulsion towards the aqueous external phase, also known as the treat ratio, also influences the extraction performance. The extraction efficiency increases with the treat ratio [31]. According to Ooi *et al.* [66], the extraction rate is enhanced due to the improvement of solute permeation and stripping. The molar ratio of internal stripping agents to the solute in the external phase increases with emulsion volume, therefore, increasing the capacity of the internal reagent for solute recovery [67]. On the other hand, Kumbasar [68] found that the increment of the treat ratio beyond a certain level could affect emulsion stability owing to the creation of a viscous emulsion.

Extraction time or agitation time is also another parameter affecting the ELM process. It exemplified the contact period amid the emulsion and the aqueous external phase during the ELM process. Sufficient time is required for the distribution of a higher number of emulsion globules in the aqueous external phase. In contrast, Kulkarni [69] found that prolonged extraction time induced water transportation from the aqueous external phase toward the emulsion phase, resulting in the swelling of the emulsion. Consequently, the accumulation of water in the internal phase will cause emulsion breakage and result in internal phase leakage [70]. Hence, a longer extraction time is not recommended, but adequate time is compulsory for the emulsion dispersion and extraction process.

Meanwhile, in the case of carrier-mediated ELM transport, the pH of the external aqueous phase is critical for solute extraction. This is on account of the carrier being controlled by the pH of the feed phase solution to attain equilibrium during the reaction with the targeted solute. The external phase pH also needs to be adjusted to prevent emulsion swelling and ensure emulsion stability [71].

Notably, the stability of the emulsion should be taken into account for a long-term operation. Emulsion stability is designated as the capability to prevent globule breakage or leakage of extracted solute during the ELM process [72]. During the emulsion dispersion in the aqueous feed phase, adequate stability is essential to extract the targeted solute into the internal stripping phase. The breakage of the emulsion will lower the extraction efficiency, and the extracted solute will be lost. To recover the extracted solute, the emulsion should be easily demulsified by the use of appropriate demulsification methods, such as heating, chemical demulsification, or applying an electrostatic field.

Emulsion breakage or rupture of the emulsion can reduce the extracted solute concentration. As a result, the volume of the stripping phase affects the solute recovery. Meanwhile, swelling involves the transportation of some aqueous feed phase into the emulsion phase. This drives the increase in stripping phase volume, causing emulsion stability problems during the process. There are several studies on factors contributing to emulsion stability that are particularly related to the emulsion formulation, processing condition, and emulsification procedure.

3.5 Conclusion

In this study, the improvement of extraction performance through the employment of synergists in several metal extraction processes was demonstrated. Besides, the incorporation of vegetable oil as an environmentally friendly material in the organic liquid formulation has been presented. The results show that 98% of Ag extraction was synergistically obtained by employing 0.0002 M Cyanex 302 + 0.0003 M Cyanex 272 in palm oil with a synergistic factor of 62.70. Almost 99% of Cr (VI) extraction was achieved using 0.01 M Cyanex 272 + 0.04 M Cyanex 302 with a synergistic factor of 7.73. Meanwhile, 83% of nickel ion extraction was obtained through the combination of 0.08 M LIX 63 with 0.02 M D2EHPA in palm oil, with a 29.56 synergistic factor value. On the other hand, 80% of Zn was extracted using 1 mM Cyanex 302 + 9 mM D2EHPA with a 2.90 synergistic coefficient value. It is inferred that the developed formulation can enhance the extraction using ELM as one of the sustainable methods for various metal removals in the future.

Acknowledgment

The fundings of this research were provided by the Ministry of Higher Education (MOHE), Universiti Teknologi Malaysia (Research Grant; PDRU: Q.J130000.21A2.05E51), CRG 40.0: Q.J130000.2451.08G02, and CRG 40.4: R.J130000.7351.4B425), and Centre of Lipids Engineering and Applied Research (CLEAR).

References

1. Niazi, N.K., Murtaza, B. *et al.*, Removal and recovery of metals by biosorbents and biochars derived from biowastes, M.N.V. Prasad and K. Shih (Eds.), pp. 149–177, Academic Press, United States, 2016.
2. Krishnan, S., Zulkapli, N.S., Kamyab, H., Taib, S.M., Din, M.F.B.M., Majid, Z.A., Chaiprapat, S., Kenzo, I., Ichikawa, Y., Nasrullah, M., Chelliapan, S., Othman, N., Current technologies for recovery of metals from industrial wastes: An overview. *Environ. Technol. Innov.*, 22, 101525, 2021.
3. Kumar, M., Nandi, M., Pakshirajan, K., Recent advances in heavy metal recovery from wastewater by biogenic sulfide precipitation. *J. Environ. Manage.*, 278, 111555, 2021.

4. Noah, N.F.M., Sulaiman, R.N.R., Othman, N., Jusoh, N., Rosly, M.B., Extractive continuous extractor for chromium recovery: Chromium (VI) reduction to chromium (III) in sustainable emulsion liquid membrane process. *J. Clean. Prod.*, 247, 119167, 2020.
5. Jusoh, N., Rosly, M.B., Othman, N., Rahman, H.A., Noah, N.F.M., Sulaiman, R.N.R., Selective extraction and recovery of polyphenols from palm oil mill sterilization condensate using emulsion liquid membrane process. *Environ. Sci. Pollut. Res.*, 27, 23246, 2020.
6. Guimarães, A.S., Silva, L.A., Pereira, A.M., Correia, J.C.G., Mansur, M.B., Purification of concentrated nickel sulfuric liquors via synergistic solvent extraction of calcium and magnesium using mixtures of D2EHPA and cyanex 272. *Sep. Purif. Technol.*, 239, 116570, 2020.
7. Wu, M., He, H., Xu, F., Xu, Z., Zhang, W., He, Z., Qu, J., Chi, R., Huang, L., High-efficient and selective extraction of hf over Zr with DIBK-P350 synergistic extraction system. *Sep. Purif. Technol.*, 212, 255, 2019.
8. Belkhouche, N.E., Amine Didi, M., Villemin, D., Separation of nickel and copper by solvent extraction using di-2 ethylhexylphosphoric acid-based synergistic mixture. *Solvent Extr. Ion Exch.*, 23, 677, 2005.
9. Andrade, C., Arnold, L., Motabar, D., Aspelund, M., Tang, A., Hunter, A., Chung, W.K., An integrated approach to aggregate control for therapeutic bispecific antibodies using an improved three column mab platform-like purification process. *Biotechnol. Prog.*, 35, e2720, 2019.
10. Singh, S., Dhami, P., Tripathi, S., Dakshinamoorthy, A., Studies on the recovery of uranium from phosphoric acid medium using synergistic mixture of (2ethyl hexyl) Phosphonic acid, mono (2-ethyl hexyl) ester (PC88A) and tri n-butyl phosphate (TBP). *Hydrometallurgy*, 95, 170, 2009.
11. Fouad, E.A., Separation of copper from aqueous sulfate solutions by mixtures of Cyanex 301 and LIX® 984N. *J. Hazard. Mater.*, 166, 720, 2009.
12. Chang, S.H., Utilization of green organic solvents in solvent extraction and liquid membrane for sustainable wastewater treatment and resource recovery—A review. *Environ. Sci. Pollut. Res.*, 27, 32371, 2020.
13. Muthuraman, G. and Palanivelu, K., Transport of textile dye in vegetable oils based supported liquid membrane. *Dyes Pigments*, 70, 99, 2006.
14. Mahmoud, A.S., Ghaly, A.E., Brooks, M.S., Removal of dye from textile wastewater using plant oils under different pH and temperature conditions. *Am. J. Environ. Sci.*, 3, 205, 2007.
15. Chakrabarty, K., Saha, P., Ghoshal, A.K., Separation of mercury from its aqueous solution through supported liquid membrane using environmentally benign diluent. *J. Membr. Sci.*, 350, 395, 2010.
16. Björkegren, S., Fassihi Karimi, R., Martinelli, A., Jayakumar, N.S., Hashim, M.A., A new emulsion liquid membrane based on a palm oil for the extraction of heavy metals. *Membranes*, 5, 168, 2015.

17. Noah, N.F.M., Othman, N., Jusoh, N., Highly selective transport of palladium from electroplating wastewater using emulsion liquid membrane process. *J. Taiwan Inst. Chem. Eng.*, 64, 134, 2016.
18. Park, Y., Skelland, A.H.P., Forney, L.J., Kim, J.H., Removal of phenol and substituted phenols by newly developed emulsion liquid membrane process. *Water. Res.*, 40, 1763, 2006.
19. Baker, R.W., *Membrnae Technology and Applications*, McGraw-Hill, United Kingdom, 2000.
20. Ho, W.S. and Kamalesh, K.S., *Membrane Handbook*, Chapman & Hall, New York, 1992.
21. Villora, G., Liquid membranes, supported and emulsion, E.M. Hoek and V.V. Tarabara (Eds.), John Wiley & Sons, United States, 2013.
22. Kargari, A., Kaghazchi, T., Soleimani, M., Mathematical modeling of emulsion liquid membrane pertraction of gold (III) from aqueous solutions. *J. Membr. Sci.*, 279, 380–388, 2006.
23. Hiroshi, T., *Liquid Membranes: Chemical Applications*, CRC Press, Boca Raton, Florida, 1990.
24. Kumar, A. and Thakur, A., Parametric optimization of green synergistic reactive extraction of lactic acid using trioctylamine, aliquat336, and butan-2-ol in sunflower oil by response surface methodology. *Chem. Eng. Commun.*, 206, 1072, 2018.
25. Rahman, H.A., Jusoh, N., Othman, N., Rosly, M.B., Sulaiman, R.N.R., Noah, N.F.M., Green formulation for synthetic dye extraction using synergistic mixture of acid-base extractant. *Sep. Purif. Technol.*, 209, 293, 2019.
26. Sulaiman, R.N.R. and Othman, N., Removal and recovery of chromium(VI) ion via tri-n-octyl methylammonium chloride-kerosene polypropylene supported liquid membrane. *Malaysian J. Anal. Sci.*, 21, 416, 2017.
27. Sulaiman, R.N.R., Othman, N., Amin, N.A.S., Emulsion liquid membrane stability in the extraction of ionized nanosilver from wash water. *J. Ind. Eng. Chem.*, 20, 3243, 2014.
28. Kiani, S. and Mousavi, S.M., Ultrasound assisted preparation of water in oil emulsions and their application in arsenic (V) removal from water in an emulsion liquid membrane process. *Ultrason. Sonochem.*, 20, 373, 2013.
29. Benyahia, N., Belkhouche, N., Jönsson, J.Å., A comparative study of experimental optimization and response surface methodology of Bi(III) extraction by emulsion organophosphorus liquid membrane. *J. Environ. Chem. Eng.*, 2, 1756, 2014.
30. Mokhtari, B. and Pourabdollah, K., Emulsion liquid membrane for selective extraction of Bi(III). *Chin. J. Chem. Eng.*, 23, 641, 2015.
31. Goyal, R.K., Jayakumar, N.S., Hashim, M.A., Chromium removal by emulsion liquid membrane using [BMIM]+[NTf2]–as stabilizer and TOMAC as extractant. *Desalination*, 278, 50, 2011.
32. Zeng, L., Zhang, Y., Liu, Q., Yang, L., Xiao, J., Liu, X., Yang, Y., Determination of mass transfer coefficient for continuous removal of cadmium by emulsion

liquid membrane in a modified rotating disc contactor. *Chem. Eng. J.*, 289, 452, 2016.

33. Kumbasar, R.A., Selective extraction of cobalt from strong acidic solutions containing cobalt and nickel through emulsion liquid membrane using TIOA as carrier. *J. Ind. Eng. Chem.*, 18, 2076, 2012.
34. Sulaiman, R.N.R., Othman, N., Amin, N.A.S., Recovery of ionized nanosilver from wash water solution using emulsion liquid membrane process. *J. Teknol. (Sci. Eng.)*, 65, 33, 2013.
35. Eyupoglu, V. and Kumbasar, R.A., Extraction of Ni(II) from spent Cr–Ni electroplating bath solutions using LIX 63 and 2BDA as carriers by emulsion liquid membrane technique. *J. Ind. Eng. Chem.*, 21, 303, 2015.
36. Lende, A.B. and Kulkarni, P.S., Selective recovery of tungsten from printed circuit board recycling unit wastewater by using emulsion liquid membrane process. *J. Water Process Eng.*, 8, 75, 2015.
37. Othman, N., Noah, N.F.M., Poh, K.W., Yi, O.Z., High performance of chromium recovery from aqueous waste solution using mixture of palm-oil in emulsion liquid membrane. *Procedia Eng.*, 148, 765, 2016.
38. Nguyen, T.H. and Hoa, N.V.N., Review on the comparison of the chemical reactivity of Cyanex 272, Cyanex 301 and Cyanex 302 for their application to metal separation from acid media. *Metals*, 10, 1, 2020.
39. Noah, N.F.M. and Othman, N., Emulsion stability of palladium extraction containing Cyanex 302 as a mobile carrier in emulsion liquid membrane process. *Chem. Eng. Trans.*, 56, 1069, 2017.
40. Sole, K.C. and Hiskey, J.B., Solvent extraction of copper by Cyanex 272, Cyanex 302 and Cyanex 301. *Hydrometallurgy*, 37, 129, 1995.
41. El-Hefny, N.E. and Daoud, J.A., Extraction of copper(II) by Cyanex 302 in kerosene from different aqueous media. *Solvent Extr. Ion Exch.*, 25, 831, 2007.
42. Nguyen, V.N.H., Nguyen, T.H., Lee, M.S., Review on the comparison of the chemical reactivity of Cyanex 272, Cyanex 301 and Cyanex 302 for their application to metal separation from acid media. *Metals*, 10, 1105, 2020.
43. Chauhan, S. and Patel, T., A review on solvent extraction of nickel. *Int. J. Eng. Res. Technol.*, 3, 1315, 2014.
44. Cerpa, A. and Alguacil, F.J., Separation of cobalt and nickel from acidic sulfate solutions using mixtures of di(2-ethylhexyl)phosphoric acid (DP-8R) and hydroxyoxime (ACORGA M5640). *J. Chem. Technol. Biotechnol.*, 79, 455, 2004.
45. De Souza, M.F.A. and Mansur, M.B., Competing solvent extraction of calcium and/or nickel with Cyanex 272 and/or d2ehpa. *Braz. J. Chem. Eng.*, 36, 541, 2019.
46. Jusoh, N., Othman, N., Sulaiman, R.N.R., Noah, N.F.M., Kamarudin, K.S.N., Zaini, M.A.A., Sidik, D.A.B., Development of palm oil-based synergist liquid membrane formulation for silver recovery from aqueous solution. *J. Membr. Sci. Res.*, 7, 59, 2021.

47. Ichlas, Z.T. and Ibana, D.C., Process development for the direct solvent extraction of nickel and cobalt from nitrate solution: Aluminum, cobalt, and nickel separation using Cyanex 272. *Int. J. Miner. Metall. Mater.*, 24, 37, 2017.
48. Kumbasar, R.A., Selective separation of chromium (VI) from acidic solutions containing various metal ions through emulsion liquid membrane using trioctylamine as extractant. *Sep. Purif. Technol.*, 64, 56, 2008.
49. Liu, Y., Lee, M.S., Jeon, H.S., Solvent extraction of pr and nd from chloride solution by mixtures of acidic extractants and LIX 63. *J. Korean Inst. Metals Mater.*, 54, 592, 2016.
50. Cheng, C.Y., Barnard, K.R., Zhang, W., Robinson, D.J., Synergistic solvent extraction of nickel and cobalt: A review of recent developments. *Solvent Extr. Ion Exch.*, 29, 719, 2011.
51. Sulaiman, R.N.R., Noah, N.F.M., Othman, N., Jusoh, N., Rosly, M.B., Synergetic formulation of Cyanex 272/Cyanex 302 for hexavalent chromium removal from electroplating wastewater. *Korean J. Chem. Eng.*, 38, 514, 2021.
52. Rajewski, J. and Religa, P., Synergistic extraction and separation of chromium(III) from acidic solution with a double-carrier supported liquid membrane. *J. Mol. Liq.*, 218, 309, 2016.
53. Andrade, F. and Elizalde, M.P., Synergistic extraction of ni(ii) by mixtures of lix 860 and bis(2-ethylhexyl) phosphoric acid. *Solvent Extr. Ion Exch.*, 23, 85, 2005.
54. Sulaiman, R.N.R. and Othman, N., Synergistic green extraction of nickel ions from electroplating waste via mixtures of chelating and organophosphorus carrier. *J. Hazard. Mater.*, 340, 77, 2017.
55. Zereshki, S., Shokri, A., Karimi, A., Application of a green emulsion liquid membrane for removing copper from contaminated aqueous solution: Extraction, stability, and breakage study using response surface methodology. *J. Mol. Liq.*, 325, 115251, 2021.
56. Othman, N., Noah, N.F.M., Poh, K.W., Yi, O.Z., High performance of chromium recovery from aqueous waste solution using mixture of palm-oil in emulsion liquid membrane. *Procedia Eng.*, 148, 765, 2016.
57. Rosly, M.B., Jusoh, N., Othman, N., Rahman, H.A., Noah, N.F.M., Sulaiman, R.N.R., Effect and optimization parameters of phenol removal in emulsion liquid membrane process via fractional-factorial design. *Chem. Eng. Res. Des.*, 145, 268, 2019.
58. Barad, J.M., Chakraborty, M., Bart, H.J., Stability and performance study of water-in-oil-in-water emulsion: Extraction of aromatic amines. *Ind. Eng. Chem. Res.*, 49, 5808, 2010.
59. Kislik, V.S., Preface, V.S. Kislik (Ed.), pp. xiii–xiv, Elsevier, United Kingdom, 2010.
60. Othman, N., Noah, N.F.M., Shu, L.Y., Ooi, Z.Y., Jusoh, N., Idroas, M., Goto, M., Easy removing of phenol from wastewater using vegetable oil-based organic solvent in emulsion liquid membrane process. *Chin. J. Chem. Eng.*, 25, 45, 2017.

61. Mortaheb, H.R., Amini, M.H., Sadeghian, F., Mokhtarani, B., Daneshyar, H., Study on a new surfactant for removal of phenol from wastewater by emulsion liquid membrane. *J. Hazard. Mater.*, 160, 582, 2008.
62. Kumbasar, R.A., Studies on extraction of chromium (VI) from acidic solutions containing various metal ions by emulsion liquid membrane using alamine 336 as extractant. *J. Membr. Sci.*, 325, 460, 2008.
63. Jusoh, N., Othman, N., Rosly, M.B., Extraction and recovery of organic compounds from aqueous solution using emulsion liquid membrane process. *Mater. Today Proc.*, 47, 1301–1306, 2021.
64. Sulaiman, R.N.R., Rahman, H.A., Othman, N., Rosly, M.B., Jusoh, N., Noah, N.F.M., Extraction of reactive dye via synergistic aliquat 336/D2EHPA using emulsion liquid membrane system. *Korean J. Chem. Eng.*, 37, 141, 2020.
65. Hasan, M.A., Selim, Y.T., Mohamed, K.M., Removal of chromium from aqueous waste solution using liquid emulsion membrane. *J. Hazard. Mater.*, 168, 1537, 2009.
66. Ooi, Z.Y., Harruddin, N., Othman, N., Recovery of kraft lignin from pulping wastewater via emulsion liquid membrane process. *Biotechnol. Prog.*, 31, 1305, 2015.
67. Abbassian, K. and Kargari, A., Modification of membrane formulation for stabilization of emulsion liquid membrane for extraction of phenol from aqueous solutions. *J. Environ. Chem. Eng.*, 4, 3926, 2016.
68. Kumbasar, R.A., Selective extraction of chromium (VI) from multicomponent acidic solutions by emulsion liquid membranes using tributhylphosphate as carrier. *J. Hazard. Mater.*, 178, 875, 2010.
69. Kulkarni, P.S., Recovery of uranium (VI) from acidic wastes using tri-n-octylphosphine oxide and sodium carbonate based liquid membrane. *Chem. Eng. J.*, 92, 209, 2003.
70. Chiha, M., Hamdaoui, O., Ahmedchekkat, F., Pétrier, C., Study on ultrasonically assisted emulsification and recovery of copper(II) from wastewater using an emulsion liquid membrane process. *Ultrason. Sonochem.*, 17, 318, 2010.
71. Malik, M.A., Hashim, M.A., Nabi, F., Extraction of metal ions by elm separation technology. *J. Dispers. Sci. Technol.*, 33, 346, 2011.
72. Moyo, F. and Tandlich, R., Mini-review on the use of liquid membranes in the extraction of platinum group metals from mining and metal refinery wastewaters/side-streams. *J. Bioremediat. Biodegrad.*, 5, 228, 2014.

4

Chemical Activation of Carbonized Neem Seed as an Effective Adsorbent for Rhodamine B Dye Adsorption

Edwin Andrew Ofudje[1]*, Samson O. Alayande[2], Abimbola A. Ogundiran[3], Ezekiel Folorunso Sodiya[1], Oyesolape Basirat Akinsipo-Oyelaja[3], Godswill Akhigbe[4] and Olugbenga Bowale Oladeji[5]

[1]*Department of Chemical Sciences, Mountain Top University, Ogun State, Nigeria*
[2]*Department of Industrial Chemistry, First Technical University, Ibadan, Oyo State, Nigeria*
[3]*Department of Chemical Sciences, Tai Solarin University of Education, Ijebu-Ode, Ogun State, Nigeria*
[4]*Department of Chemical Sciences, McPherson University, Ogun State, Nigeria*
[5]*Department of Chemistry, Federal University of Agriculture, Abeokuta, Ogun State, Nigeria*

Abstract

The behavior pattern of rhodamine B dye adsorption from wastewater on chemically activated carbonized neem seed powder was considered via the batch adsorption process. The surface chemistry of neem seed biomass was investigated using a scanning electron microscope (SEM), X-ray diffraction (XRD), and Fourier transform infrared spectroscopy (FT-IR). Various parameters, such as dye concentration, adsorbent amount, temperature, contact time, and solution pH, were tested. The pseudo-first-order kinetic model demonstrated good suitability with better correlation coefficients and adsorption capacities for raw and carbonized neem seed biomass, while that of the magnetic carbonized neem seed obeyed the second-order model. The maximum adsorption capacities from the Langmuir isotherm for raw neem seed (RNS), carbonized neem seed (CNS), and magnetized carbonized neem seed (MCNS) are 106.226, 130.438, and 188.532 mg·g^{-1}, respectively. The heats evolved by RNS, CNS, and MCNS are 7.488 kJ·mol^{-1}, 7.148 kJ·mol^{-1}, and 23.570 kJ·mol^{-1}, respectively. It was deduced that although both

**Corresponding author*: eaofudje@mtu.edu.ng; ofudjeandrew4real@yahoo.com

Shahid-ul-Islam, Abid Hussain Shalla and Mohammad Shahadat (eds.) Green Chemistry for Sustainable Water Purification, (79–106) © 2023 Scrivener Publishing LLC

the RNS and CNS showed good adsorption potency toward RB dye, the MCNS demonstrated better affinity towards the elimination of RB, thus offering the advantage of magnetic separation and eliminating the challenges connected with the removal of adsorbents from aqueous media.

***Keywords*:** Adsorption, characterization, neem seed, pollution, rhodamine B

4.1 Introduction

The rapid increase in population, expansion in social civilization coupled with recent development in industrial and technological growth has resulted in a spiky increase in innovation and municipal growth [1]. This spiky increase in industrial activities has increased the spate of destruction of the ecosystems and equally caused a serious threat to human health [2]. Nowadays, thousands of dyes are prepared for printing and dyeing industries while their waste products are been discharged directly into the water stream. Large quantities of dye are often generated from the textile industry which contaminates waste systems [3]. Dyestuffs in effluents affect the photosynthesis process because of their color and the fact that they cannot be broken down by biological means [4–7], could be potentially toxic, carcinogenic, or mutagenic [4–7]. Water-borne diseases arising from the breeding of bacteria, viruses, and vectors from such dirty water could also be a serious environmental challenge [8]. The Rhodamine B (RB) is commonly utilized as a pigment in paint, textile, and paper industries [9], which has been documented to be mutagenic and causes reproductive damages [10, 11]. As such, the search for possible ways of mitigating against the discharge of contaminants by various industries becomes imperative. Conventional methods of treating wastewater had been reported not only to be very expensive but also not effective [12]. The use of adsorbents is thus an alternative, striking and ecofriendly practice for decontaminating dyes in wastewater and is fairly economical when match up with other treatment methods [13, 14]. The dominance of adsorption in the environmental chemistry for the treatment of wastewater has been causing great attention, as a result of its low price and ease of operation [15]. Activated carbon (AC) is considered an excellent adsorbent as a result of its great permeable structure, and low acid/base reactivity and could be useful in the elimination of vast organic and inorganic pollutants [16, 17]. However, the cost of obtaining pure commercial activated carbon is very expensive coupled with the loss of adsorption competence after rebirth are some of the inconveniences which are normally come across during separation [18]. Conversely, separation via the magnetic process has been proposed as a more favored technique in the deletion of contaminants due to its ease of operation, suitability for the heterogeneous systems in bulk solutions, and a very quick separation technique orchestrated by an external magnetic field [18]. Thus, the needs for low-cost and effective magnetic activated carbon from waste materials become imperative. In this study, the preparation, characterization, and Rhodamine B adsorption onto magnetic activated carbon prepared from neem seed under different operating conditions via batch adsorption process were evaluated. Operating factors like initial RB concentration, hydrogen ion concentration, activated carbon dose, contact time, and temperature were assessed. Isotherms and kinetics studies for both adsorbents were established.

4.2 Materials and Methods

4.2.1 Chemicals

Rhodamine B (RB) ($C_{28}H_{31}CN_2O_3$), with 95% purity (Sigma-Aldrich, USA) was used without further purification. Deionized water was used to dissolve 1 g of RB in other to prepare a stock solution of RB, while several RB concentrations were made through the dilution process. Measurements of color changes were performed via UV-vis-spectrophotometer at a wavelength of maximum absorption (λmax) of 555 nm.

4.2.2 Preparation of Adsorbent

Samples of raw neem seed (RNS) were sourced locally from Foursquare Camp, Ajebo, Nigeria, and reduced to small pieces, while external dirt was removed via washing with distilled water. Samples were air-dried for 5 days, followed by drying at 110 °C for 12 hours in an oven and thereafter pulverized. In other to prepare the activated carbon from the dried neem seed powder, about 200 g of the dried samples were carbonized at 800 °C under N_2 gas flow (100 cm^3/min) in a muffle furnace. After reaching 800 °C, the heating process was maintained for 25 min and thereafter cooled at room temperature under N_2 flow (100 cm^3/min). This was referred to as carbonized neem seed (CNS). Figure 4.1 shows the molecular structure of Rhodamine B dye and raw neem seeds in form of beads.

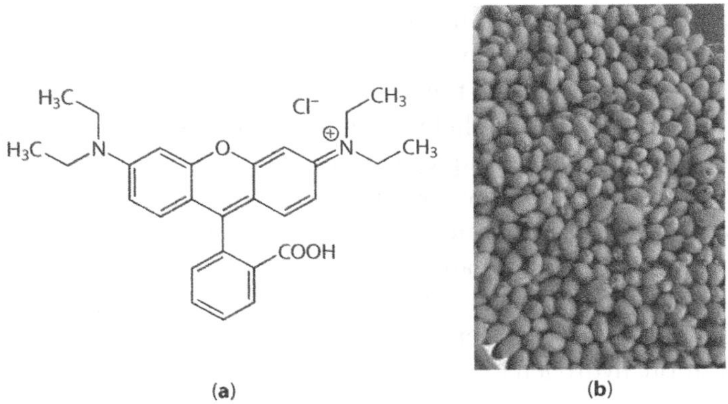

Figure 4.1 (a) chemical structure of Rhodamine B dye and (b) neem seed.

4.2.3 Magnetic Activation Carbonized Neem Seed

In other to prepare the magnetic form of the carbonized neem seed, 50 g of CNS was dissolved in a magnetic solution containing the mixture of $FeCl_3$ and $FeSO_4$ in a 500 ml beaker and stirred with the aid of a magnetic stirrer. Solution pH between 10 and 11 was maintained using NaOH solution and the process was left to age for 24 h, separated, and cleansed with distilled water. The obtained product was allowed to dry and this was referred to as magnetic carbonized neem seed (MCNS).

4.2.4 Adsorbent Characterizations

The surface chemistry of the various adsorbents was determined with different analytical tools which were done with a Hitachi (Japan) S-3000H scanning electron microscope to picture the imaging of the adsorbent surface. FT-IR spectrometry (Bruker Optics, TENSOR 27 series FT-IR spectrometer) was engaged to determine the functional groups of the adsorbents by the KBr disc method. The X-ray diffraction (XRD) were measured by PANalytical (X'Pert PRO, Netherland) in the range between $2\theta = 10–60°$ with Cu Kα monochromatic radiation (1.5406 A°). In other to measure the point of zero charges (PHP C), the modified method of Oyetade *et al.* [18] was used. Briefly, 50 cm^3 of NaCl solutions was measured into three different conical flasks and the solution pH was regulated within the pH of 1 to 10 using either HCl or NaOH solution. Then 10 g of the prepared adsorbents were introduced into each flask and the content was equilibrated on an orbital shaker for 12 h. Thereafter, the mixture was separated final pH was measured. A plot of change in pH against initial was used to obtain the pHPZC of the adsorbents by measuring the intersection point of the curve. The thermal property of RNS was investigated with the aid of SDT Q600 V8.3 Build 101simultaneous DSC-TGA instrument.

4.2.5 Batch Adsorption Experiments

The batch experiment was achieved via mixing (30 mg) of the prepared sorbent with 25 mL of RB solution in clean Erlenmeyer flasks and contacted at 150 rpm on a temperature control shaker for 100 mins to arrive at equilibrium. The medium pH was regulated with 0.1M NaOH/HCl prior to adsorbent introduction. Dye concentration was analyzed via UV-visible spectrophotometer. Operating conditions, such as effects of dosage (10–50.0 mg), contact time (10–160 min), hydrogen ion concentration (2–8), initial RB dye concentration (50–300 mg L^{-1}), and temperature (25–65 °C) were examined. The quantity of dye adsorbed at, qe (mgg^{-1}), was estimated via:

$$q_e = \frac{C_i - C_e}{m} \times V \tag{4.1}$$

where C_i is RB concentration at the beginning of the reaction (mgL^{-1}), C_e is RB concentration after the reaction has attained equilibrium (mgL^{-1}), V is RB volume utilized (L) and m is the adsorbent mass used (g).

Percentage removal of RB was evaluated as follows:

$$\text{Percentage removal} = \frac{C_i - C_e}{C_i} \times 100 \tag{4.2}$$

4.3 Results and Discussion

The physicochemical characteristics of the prepared RNS, CNS, and MCNS adsorbents are shown in Table 4.1. The surface area obtained are 566, 782, and 954 m^2/g for RNS, CNS, and MCNS respectively with the magnetic carbonized neem seed showing a significantly high value than the other two adsorbents prepared. With higher surface area, adsorbents have better adsorption capacity which enhances the adsorbents to absorb more pollutants onto the surface of the adsorbents. The chemical activated neem seed,

Table 4.1 The physicochemical characteristics of the prepared adsorbents.

Parameters	RNS	CNS	MCNS
Surface area (m^2/g)	566	782	954
Average pore size (nm)	14.21	12.44	10.36
Pore volume (cm^2/g)	0.16	0.34	0.67
Bulk density (g/cm^2)	1.46	2.37	5.88
pH_{ZPC}	4.3	4.5	4.7

MCNS gave the largest surface area, 954 m^2/g when compared with CNS and RNS which is because the activating agent has created new pores and widened the existing pores. This is evident with the highest pore volume exhibited by the magnetic carbonized neem seed (0.67 cm^3/g), followed by CNS (0.34 cm^3/g) and the least value was demonstrated by RNS (0.16 cm^3/g). The values of the average pore size and bulk density obtained are 14.21, 12.44, 10.36 (nm) and 1.46, 2.37, 5.88 (g/cm^2) for RNS, CNS, and MCNS respectively.

The XRD pattern for raw neem seed (RNS), carbonized neem seed (CNS) and magnetic carbonized neem seed (MCNS) are shown in Figure 4.2. From this figure, the effects of physical and chemical treatment on the raw neem seed can be seen. For the RNS, two main broad diffraction peaks located at $2\theta = 18.3°$ and 22.1° were exhibited depicting the presence the of amorphous structure of carbon. The result the of XRD diffractogram for CNS showed that tactivatededted carbons contain high crystallinity afteractivationion process with the appearance of many sharp peaks corresponding to carbon structure. Prominent peaks were located at around $2\theta = 28$, 31–34°, 40–41°, 50.2°, and 60.2°. Fomagneticiactivateded, the sharp peaks that were observed at 2θ equals 30.13°, 31.30°, 36.85°, 57.15°, and 63.10° were due to the presence of Fe_3O_4 (JCPDS No. 75-0033), while those observed at $2\theta = 24.40°$, 33.35°, 41.10°, 50.25° were assigned to α-Fe_2O_3 (JCPDS No. 80-2377) which were introduced into the carbonized neem seed during the chemical activation process. The raw neem seed, carbonized neem seed, and magnetic carbonized neem seed before and after the adsorption of RB were analyzed using FT-IR as depicted in Figure 4.3. As seen, the raw neem seed spectra show four major absorption peaks between 3600 and 1000 cm^{-1}. These bands correspond to the vibrations of –OH and –NH between 3445 to 3620 cm^{-1}, $-CH_2$ at 2930 cm^{-1}, C=O at 1660 to 1730 cm^{-1}, and C-O between 1009 to 1107 cm^{-1} respectively

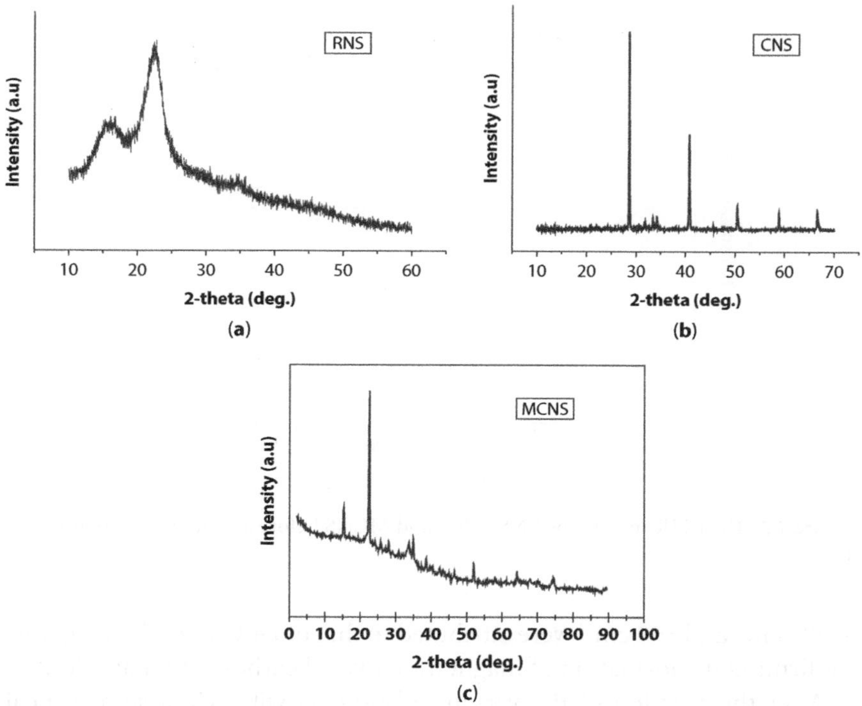

Figure 4.2 XRD spectra of (a) raw neem seed and (b) carbonized neem seed and (c) magnetic carbonized neem seed.

[13, 14]. The presence of these peaks confirmed that the neem seed is made up of lots of cellulose and hemicelluloses [14]. The FT-IR spectra of the carbonized neem seed revealed that the absorption peaks corresponding to –OH and –NH, –CH_2, and C=O in the raw neem seed had all disappeared. This is an indication of structural destruction of the cellulose and hemicelluloses molecular chain of the neem seed as the functional groups present on the raw biomass were released as volatile materials. The main absorption bands observed for the carbonized neem seed were observed between 870 to 920 and 1000 to 1104 cm^{-1} respectively and signal that carbonate functional group (C-O) is present in the structure. Analysis of the FT-IR of MCNS shows that the –OH group which was eliminated by the heat treatment has been reintroduced. After chemical activation, the prepared magnetic carbon show vibration peaks of –OH between 3450 to 3565 cm^{-1} and at 1508 cm^{-1} which could be as a result of the water molecules adsorbed, while that at 1140 cm^{-1} is for C-O stretch. The Peaks observed in the region

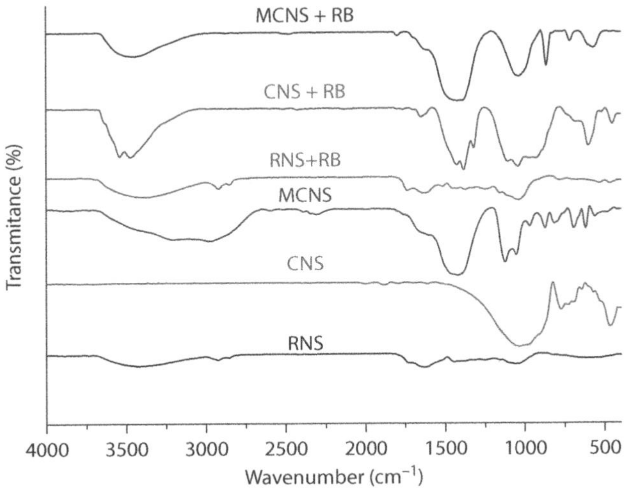

Figure 4.3 The FTIR spectra for RNS, CNS, and MCNS before and after adsorption of RB.

of 887 cm^{-1}and 569 cm^{-1} were attributed to that of Fe-O bond [18, 19], thus confirming the formation of magnetic activated carbon nanoparticles.

After the reaction of the various adsorbents with RB, a shift in peak positions coupled with the formation of new absorption bands was seen. For the RNS loaded with RB, a shift in wavenumber position for –OH, C=O, and -C-O to 3560 cm^{-1}, 1735 cm^{-1}, and 1025 cm^{-1} were observed, while the formation of new peaks for –OH and C=O at 3450 and 1680 cm^{-1} were observed respectively for the FT-IR spectra of CNS with RB load. Finally, peak position corresponding to –OH, -C-O, Fe-O shifted to 3660 cm^{-1}, 1156 cm^{-1}, and 890 cm^{-1} [18]. The shift in peak positions and the formation of new peaks (in the case of CNS) implies that these functional groups were used during the adsorption of RB by these adsorbents.

Figure 4.4 shows the scanning electron micrographs of RNS (Figure 4.4a), CNS (Figure 4.4b), and MCNS (Figure 4.4c). The surface images of RNS and CNS show rough morphology with no visible openings, whereas, MCNS demonstrates some openings with different sizes. The structure of the CNS showed a more agglomerated structure, which could have been fused by the heat treatment. The presence of the pores is due to the chemical treatment, which could serve as transport pores for the RB molecules. Upon interaction with RB, the pores previously seen on the surface of CNS disappeared, which could be as a result of the covering of the surface by the molecules RB molecules which adhered to the sorbent surface.

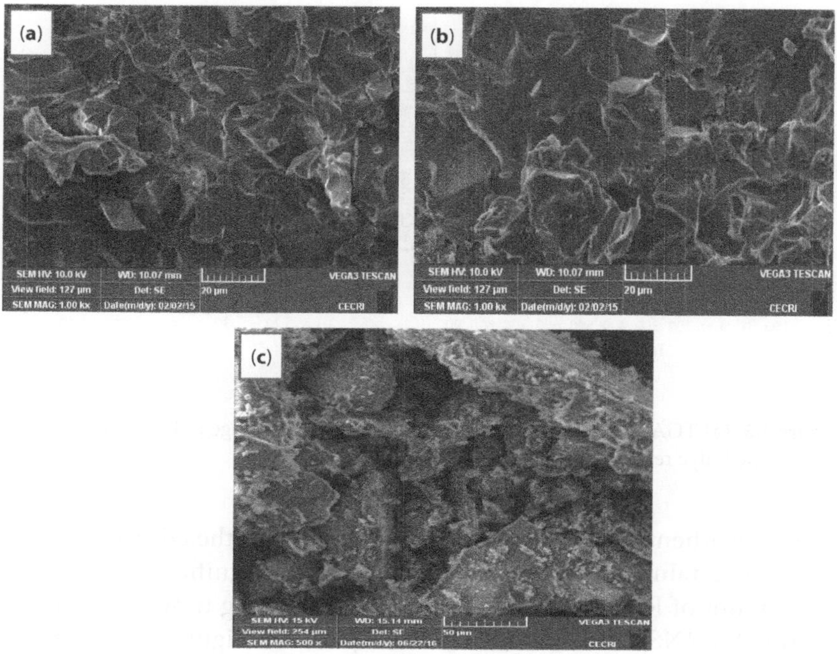

Figure 4.4 SEM images of (a) raw neem seed and (b) carbonized neem seed and (c) magnetic carbonized neem seed.

The thermal property of the raw adsorbent is shown in Figure 4.5a indicating three major weight losses of 96.63 %, 94.41 %, and 88.23% at a temperature of 101.5°C, 510°C, and 730°C, respectively [20]. The first weight loss is assigned to water molecules, the other reduction in weight corresponds to thermal degradation of cellulose components of the neem seed and the last weight is associated with the degradation of lignin components of the sample [20].

4.3.1 Adsorption Studies

The adsorption amount of RB by RNS, CNS, and MCNS increased with the dosage of the adsorbents as revealed in Figure 4.5b. The percentage removal of RB by RNS increased from 47.33% at the initial dosage of 10 mg to 66.40% at a dosage of 50 mg, while that of CNS showed an increase from 53.42% at 10 mg to 73.29% at 50 mg of adsorbent. Upon chemical treatment, the amount removal of RB increased from 57.45% at 10 mg to 85.89% at 50 mg of MCNS. The surface area, as well as the effectiveness

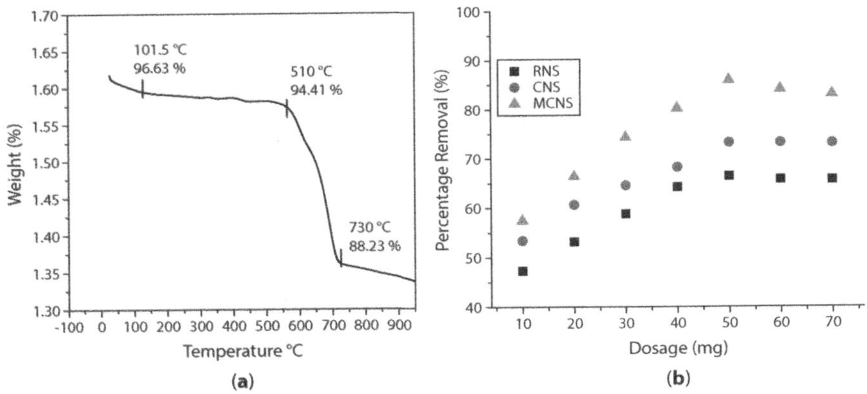

Figure 4.5 (a) TGA of raw neem seed (b) effect of biomass dosage n the amount of rhodamine B dye removal.

of the adsorbents to adsorb the RB, increased with the adsorbent dosage within a certain range, after which there was no significant difference in the amount of RB adsorbed [21]. The plots showing time-concentration relation for RNS, CNS, and MCNS are depicted in Figure 4.6, which indicated that the QE for the removal of RB by RNS increases from 4.5 mg/g at RB initial concentration of 50 mg/L to 40.22 mg at RB final concentration of 300 mg/L at a time of just 10 min. However, when the contact time was adjusted to 100 min, the adsorption capacity increased from 19.56 mg at 10 mg/L of RB to 96.78 mg for 300 mg/L of RB (Figure 4.6b). When CNS was used, the adsorption capacity rise from 7.89 mg at 10 mg/L of RB to 46.34 mg at 300 mg/L of RB within 10 min of agitation. But when the time was elevated to 100 min, the adsorption capacity of CNS increased sharply from 23.45 mg at 50 mg/L of RB to 146.54 mg when 300 mg/L of RB was used (Figure 4.6b). Upon chemical activation of the carbonized neem seed, the adsorption capacity increased from 11.34 mg at 50 mg/L of RB to 55.62 mg at 300 mg/L of RB (Figure 4.6c). But when the time was raised to 100 min, the adsorption capacity increased greatly from 34.56 mg at 10 mg/L of RB to 185.34 mg at a final concentration of 300 mg/L of RB (Figure 4.6c). For all the tested adsorbents, adsorption was quick within the first 460 min due to the presence of more receptor sites that were readily accessible for the uptake of RB onto the surface of the adsorbents, leading to a quick interface between the fabricated adsorbents and the RB molecules [18, 22]. However, with a continuous adsorption process, these receptor sites become fully occupied with the

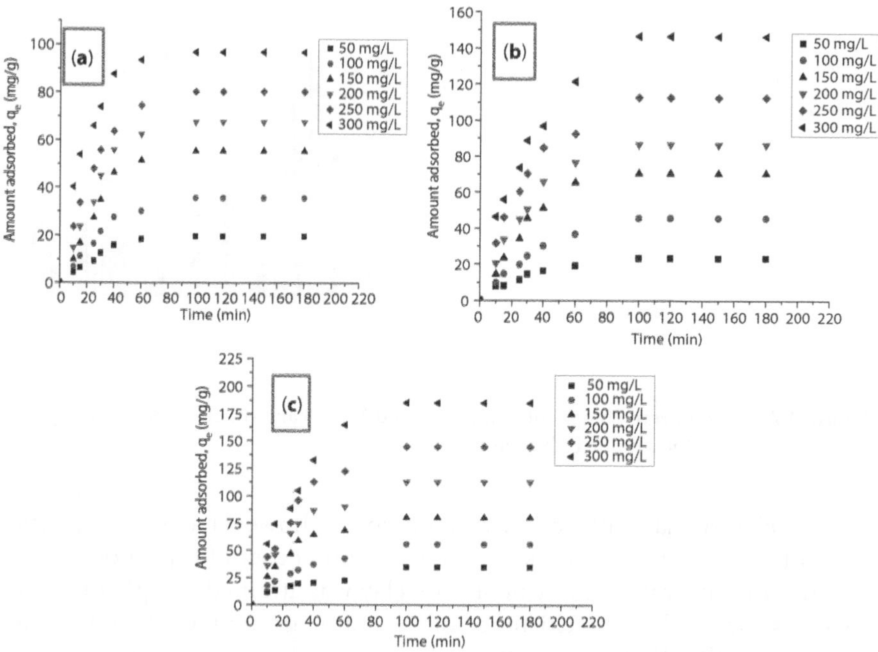

Figure 4.6 Effects of contact time and pollutant concentration on the amount of Rhodamine B dye removal.

RB molecules and as such, the rate of sorption is impeded and the movement of contaminants from the media to the adsorbent surface is lowered leading to fewer amounts of dye uptake [18, 22].

The extent of the functional group's dissociation, the adsorbent surface net charges in addition to the adsorption ability of the adsorbent are largely influenced by the pH of the solution [23]. The maximum percentage of the RB dyes removed of 73.50 %, 85.60 %, and 95.40 % was attained at pH 3.5 for RNS, CNS, and MCNS respectively (Figure 4.7b). The percentage removal thereafter decreased drastically on increasing the solution pH. The point of Zero charges of RNS, CNS, and MCNS was found to be 4.3, 4.5, and 4.7 respectively (Figure 4.7a). With a lower pH value as compared to that of the point of Zero charges of the adsorbents, the surface is expected to be positively charged and as such, the removal of cationic RB dyes would be below. But with increased pH above that of the zero point of charge, the surface becomes negatively charged thus, giving rise to better uptake of the dye.

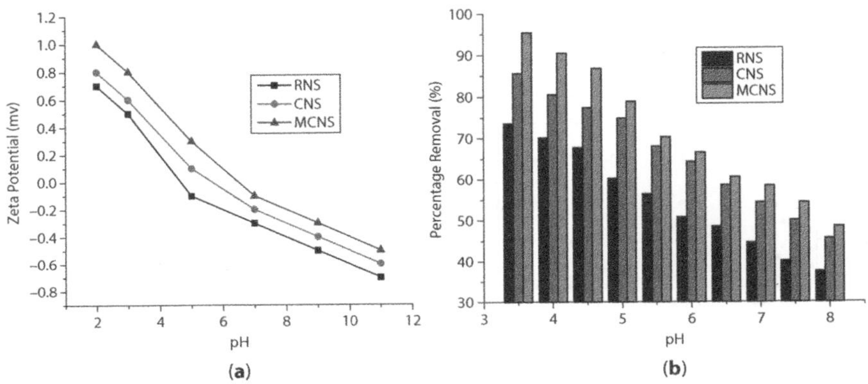

Figure 4.7 (a) point zero charges of RNS, CNS, and MCNS and (b) effect of solution pH on the amount of Rhodamine B dye removal.

This electrostatic attraction explanation is however contrary to the present findings in this work. Akeremale and Olaseni [23] reported the maximum removal of RB by bentonite clay was attained at a pH of 4 and this was ascribed to the existence of the acidic group (-COOH-) in the structure of the RB molecule. It is believed that the acidic group presents in the structure of RB increases the negative charge density of the adsorbate and as such, enhances repulsion involving the adsorbent's negatively charged surface and the molecules of dye [23]. Similarly, Kooh *et al.* [24] noted that the uptake of RB by *Casuarina equisetifolia* was optimum at a pH of 2.9 and this was as a result of the formation of both cationic and monomeric RB molecules and that at a pH higher than 4.0, RB molecules forms a dimer in solution as a result of the existence of the zwitterionic form.

4.3.2 Adsorption Kinetics of RB Dye Removal

The fitting of the sorption data of RB dye onto the three adsorbents used was achieved via Weber–Morris intra-particle diffusion model, Elovich, Pseudo-first-order (PFO), and Pseudo-second-order (PSO) using the equations described in the non-linear equations 4.3 to 4.6 [25–28]:

$$Q_t = K_{id}\, t^{0.5} + Ci \tag{4.3}$$

$$Q_t = \frac{1}{\beta}\ln(\alpha\beta * t) \tag{4.4}$$

$$Q_t = Q_e\left(1 - e^{-k_1 t}\right) \tag{4.5}$$

$$Q_t = \frac{k_2 Q_e^{\,2} t}{1 + k_2 Q_e t} \tag{4.6}$$

where qt denotes the RB concentrations (mgg^{-1}) at the time (t), and k_1(min^{-1}) and k_2($mgg^{-1}min^{-1}$) are the rate constants of the PFO and PSO respectively. The intraparticle diffusion coefficient and the diffusion constant as corresponds to the boundary layer thickness are given as k_{id}($mgg^{-1}min^{-1}$) and C_i(mgg^{-1}) respectively. From the Elovich model, the adsorption rate constant as well as the desorption rate constant which represents the accessible

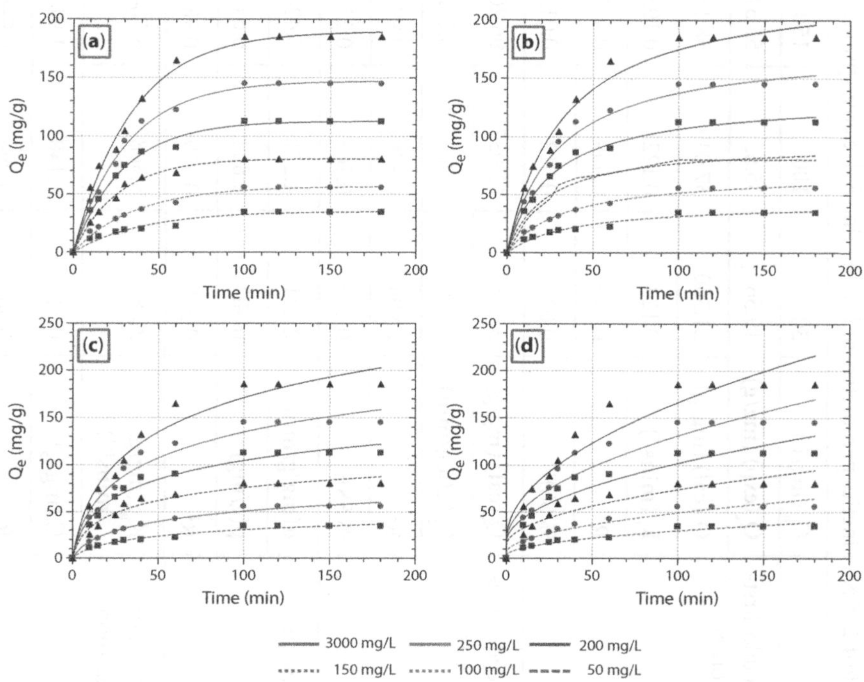

Figure 4.8 Kinetic data for the sorption of Rhodamine B onto the surface of MCNS (a) pseudo-first-order model (b) pseudo-second-order model (c) Elovich and (d) Intraparticle diffusion model fits.

Table 4.2 RNS kinetics parameters of Rhodamine B dye adsorption.

	C_o (mg/L)	50	100	150	200	250	300
Pseudo-first-order	Q_e(exp) (mg/g)	19.56	35.65	55.34	67.55	90.34	112.65
	Q_e(cal) (mg/g)	21.300	37.049	58.184	70.875	93.110	116.544
	k_1 (mins^{-1})	0.123	0.122	0.193	0.155	0.135	0.122
	R^2	0.996	0.999	0.998	0.995	0.997	0.995
	% SSE	0.027	0.012	0.015	0.015	0.009	0.010
Pseudo-first-order	Q_e(cal) (mg/g)	35.063	67.037	96.200	106.193	122.229	156.538
	k_2 (g/mg/min)	0.030	0.015	0.012	0.012	0.013	0.037
	R^2	0.994	0.998	0.995	0.997	0.989	0.993
	% SSE	0.239	0.265	0.223	0.172	0.106	0.117
Elovich	α(mg/g/mins)	2.121	4.271	5.420	7.401	9.572	12.691
	β(g/mg)	0.383	0.864	0.648	0.893	0.717	0.828
	R^2	0.988	0.979	0.985	0.997	0.996	0.998
Intraparticle Diffusion	K_{id} (mg/g/mins$^{1/2}$)	3.889	5.308	8.784	10.227	16.653	22.448
	C (mg/g)	0.312	1.148	1.915	1.155	1.820	1.900
	R^2	0.989	0.996	0.996	0.989	0.995	0.997

Table 4.3 CNS kinetics parameters of Rhodamine B dye absorption.

	C_o (mg/L)	50	100	150	200	250	300
Pseudo-first-order	Q_e(exp) (mg/g)	23.451	45.671	70.452	86.412	112.651	146.542
	Q_e(cal) (mg/g)	24.493	46.270	69.012	87.217	114.232	147.701
	k_1 (mins^{-1})	0.012	0.012	0.015	0.016	0.012	0.015
	R^2	0.998	0.999	0.998	0.997	0.998	0.995
	% SSE	0.013	0.004	0.006	0.003	0.004	0.002
Pseudo-first-order	Q_e(cal) (mg/g)	30.891	64.271	95.771	127.976	138.139	180.010
	k_2 (g/mg/min)	0.055	0.099	0.042	0.053	0.067	0.030
	R^2	0.993	0.993	0.998	0.997	0.996	0.995
	% SSE	0.096	0.123	0.108	0.145	0.068	0.069
Elovich	α(mg/g/mins)	0.362	0.934	2.331	5.316	6.658	8.283
	β(g/mg)	0.244	0.477	0.331	0.033	0.026	0.015
	R^2	0.993	0.995	0.998	0.996	0.989	0.994
Intraparticle Diffusion	K_{id} (mg/g/mins$^{1/2}$)	2.873	4.097	7.807	9.039	10.186	16.223
	C (mg/g)	0.600	0.791	0.773	0.594	3.326	8.993
	R^2	0.994	0.994	0.995	0.996	0.997	0.993

Table 4.4 MCNS kinetics parameters of Rhodamine B dye absorption.

	C_o (mg/L)	**50**	**100**	**150**	**200**	**250**	**300**
Pseudo-first-order	**Q_e(exp) (mg/g)**	34.561	55.653	80.351	112.655	144.761	185.344
	Q_e(cal) (mg/g)	21.487	46.903	70.086	98.100	104.947	134.942
	k_1 (mins^{-1})	0.223	0.417	0.683	0.879	0.957	1.540
	R^2	0.992	0.976	0.987	0.998	0.991	0.988
	% SSE	0.114	0.047	0.039	0.039	0.083	0.082
Pseudo-first-order	**Q_e(cal) (mg/g)**	32.139	56.705	82.113	114.991	149.380	188.221
	k_2 (g/mg/min)	0.083	0.012	0.022	0.018	0.048	0.026
	R^2	0.995	0.997	0.986	0.979	0.950	0.959
	% SSE	0.021	0.006	0.007	0.006	0.010	0.005
Elovich	**α(mg/g/mins)**	4.047	6.521	8.284	10.350	13.276	15.057
	β(g/mg)	0.889	0.390	0.521	0.249	0.186	0.690
	R^2	0.992	0.996	0.992	0.987	0.997	0.995
Intraparticle Diffusion	**K_{id} (mg/g/mins$^{1/2}$)**	4.986	6.823	9.948	12.349	15.130	20.392
	C (mg/g)	1.093	2.566	3.135	7.758	10.072	12.742
	R^2	0.989	0.984	0.989	0.999	0.986	0.988

spots on the sorbent surface is denoted as α (mg g^{-1} min^{-1}) and β (g mg^{-1}) respectively. The parameters of all the kinetic models were estimated from the non-linear curve as shown in Figures 4.8b–d and represented in Tables 4.2 to 4.4. The values of the equilibrium capacities (Q_e) as derived from the experimental data and the kinetic models of PFO and PSO alongside the correlation coefficients (R^2) were deployed to describe the mechanism of the three sorbents. Judging from these studies, it can be established that the kinetics data of RB adsorption by RNS and CNS fitted with the PFO indicates a physical adsorption process, while that of MCNS fitted with the PSO kinetics suggests a chemical adsorption process.

$$\%SSE = \sqrt{\frac{((Q_{(exp)} - Q_{(Cal)}) / Q_{exp})^2}{N-1}} \times 100 \tag{4.7}$$

This was further confirmed using the sum square error function (SSE) which is given in equation 4.7 with smaller values of SSE for the PFO when RNS and CNS were used and PSO for MCNS respectively. The correlation coefficients from the Elovich model were found in the range of (0.979–0.998) for RNS, (0.989–0.99) for CNS, and (0.978–0.997) for MCNS respectively thus implying the suitability of Elovich this model. The rate constant (K_{id}) of the Weber–Morris intraparticle diffusion models obtained demonstrated an increase with the initial RB concentration for all the adsorbents tested. Two distinct diffusion phases were noticed for the adsorption route with the first stage occurring rapidly within the first 40 min and the second phase proceeding until the equilibrium stage sets in.

4.3.3 Adsorption Isotherms of RB Dye Removal

Adsorption models from Langmuir isotherm [29], Freundlich isotherm [30] as well as Dubinin–Radushkevich (D-R) [31] were applied to decide the RB adsorption capacities of RNS, CNS, and MCNS adsorbents as described in equations 4.8 to 4.10 respectively. From the Langmuir isotherm, Q_{max} denotes the ceiling amount of the RB required for the formation of the entire monolayer surface, while the equilibrium constant (Lmg^{-1}) is represented as *b*.

$$Q_e = \frac{Q_{max} b C_e}{1 + b C_e} \tag{4.8}$$

$$Q_e = K_F C_e^{1/n} \tag{4.9}$$

$$Q_e = Q_s e^{-\beta\varepsilon^2} \tag{4.10}$$

The value of b can be used to obtain the separation factor (R_L) which portrays the nature of the adsorption process:

$$R_L = 1/(1 + bC_0) \tag{4.11}$$

The Freundlich isotherm constant (K_F) relates the bonding energy which defines as the distribution coefficient of the amount of RB adsorbed onto the available receptor sites the adsorbents, while the value of $1/n$ measures the surface heterogeneity of the adsorption intensity. The theoretical saturation capacity of the D-R is denoted as Q_s (mol g^{-1}), while the Polanyi potential (ε) is represented as: such that R (J mol^{-1} K^{-1}) and T (K) is the molar gas constant and temperature of the adsorption process respectively. The values of β (mol^2 J^{-2}) as related to the mean free energy E (kJ mol^{-1}) of adsorption is given as:

$$E = (2\beta)^{-0.5} \tag{4.12}$$

The nonlinear curves of *Qe* vs *Ce* as represented in Figures 4.9a–c were used to obtain the isotherm constants as given in Table 4.5. Standing on the values of the correlation coefficients (R^2) as well as the adsorption capacities of the three isotherms investigated, better suitability with the Langmuir isotherm model was observed with RNS as well as CNS (R^2 = 0.998 and 0.986), while that of the MCNS was adjudged to be better fitted with the Freundlich isotherm (R^2 = 0.987). Consequently, it can be deduced that while the uptake of RB onto the surface of RNS and CNS proceeds via homogeneous surface, that of MCNS took place on a heterogeneous surface of signifying chemical adsorption process though.

The Langmuir maximum adsorption capacities for RNS, CNS, and MCNS are 106.226, 130.438, and 188.532 mgg^{-1} respectively. The $1/n$ values obtained are 0.154, 0.225, and 0.254 RNS, CNS, and MCNS respectively thus confirming that the sorption of RB is favorable. The Qs values derived from D-R isotherms are 98.742, 120.503, and 143.362 for RNS. CNS and MCNS respectively with the magnetic-activated form showing

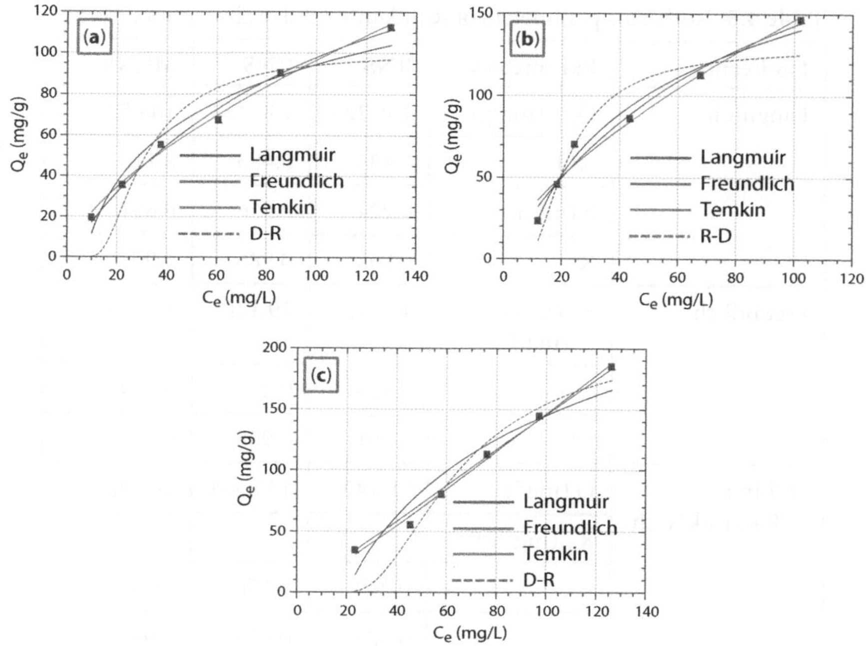

Figure 4.9 Isotherm plots for the sorption of Rhodamine B by (a) RNS, (b) CNS and (c) MCNS.

greater adsorption capacity. The values of E calculated are 0.249 for RNS, 0.136 for CNS, and 10.395 MCNS suggesting physisorption for RNS and CNS and chemisorptions for MCNS respectively. The adsorption capacities (QM) of the tested adsorbents from the Langmuir isotherm were compared with published data as indicated in Table 4.6 for the removal of RB. As illustrated in Table 4.6, the adsorption capacities of RNS, CNS, and MCNS adsorbents used in this study match up with the literature data.

4.3.4 Thermodynamic of RB Dye Removal

The impact of temperature on the elimination of RB by RNS, CNS, and MCNS can be described in terms of the relationship between the solute particles in solution and at equilibrium, as follows:

$$K_d = \frac{Q_e}{C_e} \tag{4.13}$$

Table 4.5 Isotherms parameters of Rhodamine B dye absorption.

Isotherms	Parameters	RNS	CNS	MCNS
Langmuir	Q_{max} (mg/g)	106.226	130.438	188.532
	ALL	0.135	0.485	0.113
	b (mg/L)	0.224	0.356	0.651
	R^2	0.998	0.986	0.977
Freundlich	K_F (mg/g) $(mg/L)^{-1/2}$	12.543	29.128	30.387
	1/n	0.154	0.225	0.254
	R^2	0.991	0.958	0.987
Dubinin–Radushkevich	$Q_s(mg^{-1})$	98.742	120.503	143.362
	$K_{R,}$ $(mg^{-1}L)^{1/g}$			
	E	0.249	0.136	10.395
	R^2	0.989	0.979	0.926

Table 4.6 Comparison of neem seed-derived adsorbents with literature reports.

Adsorbents	Q_{max}(mg/g)	References
Duolite C-20 resin	28.571	Al-Rashed and Al-Gaid [3]
Pomegranate peel	30.47	Rajae *et al.* [4]
CT269DR resin	164.44	Jinbei *et al.* [5]
Porous organic polymer adsorbent (DAPS-TPPOP)	59.20	Xiao-Cheng *et al.* [6]
ZSM-5 Directly Synthesized from Bangka Kaolin	128.21	Ani *et al.* [7]
Cobalt ferrite nanoparticles	5.165	Oyetade *et al.* [18]
Acid-functionalized multiwalled carbon nanotubes	42.68	Oyetade *et al.* [18]

(*Continued*)

Table 4.6 Comparison of neem seed-derived adsorbents with literature reports. (*Continued*)

Adsorbents	Q_{max} (mg/g)	References
Casuarina equisetifolia cone	82.34 mg	Kooh *et al.* [24]
Centrifuge treated Olivedos clay	58.82	Neto *et al.* [33]
Centrifuge treated Sessego clay	344.82	Neto *et al.* [33]
Cuba clay	123.46	Neto *et al.* [33]
Black tea leaves	53.2	Mohammad and Shah [34]
Modified three-dimensional layered double hydroxide	128.21	Zhu *et al.* [35]
Cassava slag biochar (HCS)	105.30	Wu *et al.* [36]
Dowex 5WX8 Resin	43.4	Khan *et al.* [37]
Zeolitic imidazolate frameworks (ZIF)	85.00	Zhang *et al.* [38]
Grass-Waste	54.00	Zahir *et al.* [39]
Raw neem seed	106.226	This study
Carbonized neem seed	130.4381	This study
Magnetic carbonized neem seed	188.532	This study

Knowledge of the equilibrium constant was deployed to ascertain the entropy change (ΔS°), free energy change (ΔG°), and the enthalpy change (ΔH°) by using the expressions [31–33]:

$$\Delta G^o = -RT \ln K_D \tag{4.14}$$

$$\ln K_d = \Delta S^o / R - \Delta H^o / RT \tag{4.15}$$

$$\Delta G^o = \Delta H^o - T\Delta S \tag{4.16}$$

From the linear plot of In *Kd* against the reciprocal of temperature, the intercept, ΔH° and slope, and ΔS° can be deduced respectively (see Figure 4.10). The spontaneous character of the process was established from the negative values observed for ΔG° which rises with the temperature, thus suggesting better diffusion of pollutant particles onto the vacant sites of the adsorbent at elevated temperature (see Table 4.7). The values of the ΔS° for RNS, CNS, and MCNS were positive suggesting a rise in the degree of randomness of the adsorption system. According to Oyetade *et al.* [18], the adsorption process could be described as entropy-driven an endothermic enthalpy is observed.

The values of ΔH° were estimated at 7.488 kJmol^{-1} for RNS, 7.148 kJmol^{-1} for CNS, and 23.570 kJmol^{-1} for MCNS. The adsorption of pollutants onto the surface of the adsorbent could proceed via physical or chemical interactions [18]. Yu and Ya-Juan [40] reported that during physical adsorption, the heat evolved is similar to that of the heat of condensation, which is between 2.1 and 20.9 kJmol^{-1}, however, that of chemical interactions is within 80 to 200 kJ mol^{-1}. Thus, RB sorption by RNS and CNS can be said to have followed physical interactions, while that of MCNS obeyed chemical interaction. Furthermore, with ΔH° being positive, the adsorption is an endothermic process (see Table 4.7). The differences observed in the reaction mechanisms of the adsorbents could be due to different chemical compositions of the adsorbents as revealed by the FT-IR analysis.

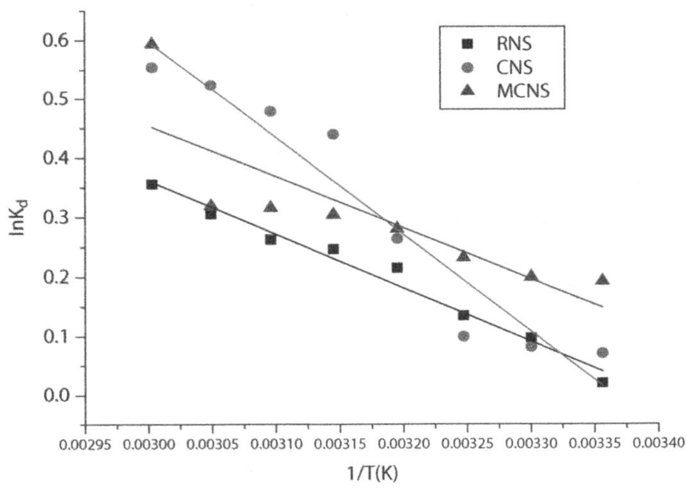

Figure 4.10 Vant Hoffe plot of *ln Kd* versus 1/T for the adsorption of Rhodamine B dye.

Table 4.7 Thermodynamic parameters of Rhodamine B dye absorption.

RNS					
T (K)	**K_d**		**ΔH $(kJmol^{-1})$**	**ΔS $(kJmol^{-1})$**	**R^2**
298	1.021	-51.534			
313	1.102	245.113			
318	1.144	344.416			
323	1,240	-559.751	7.488	25.466	0.978
328	1.280	-652.767			
333	1.300	-704.653			
338	1.357	-832.005			
CNS					
298	1.107	-174.669			
313	1.085	-205.262			
318	1.104	-253.254			
323	1.302	-686.742	7.148	25.211	0.963
328	1.552	-1160.790			
333	1.613	-1283.900			
338	1.687	-1425.940			
MCNS					
298	1.212	-176.363			
313	1.221	-503.073			
318	1.261	-593.829			
323	1.324	-730.461	23.570	45.677	0.928
328	1.356	-805.053			
333	1.371	-847.251			
338	1.376	-869.638			

4.4 Conclusions

The sorption of RB from wastewater was effectively achieved using raw neem seed, carbonized neem seed, and magnetic carbonized neem seed. The chemistry of the surface of the adsorbents was confirmed using FT-IR, XRD, and SEM respectively. It was discovered that the effectiveness of the adsorbent greatly depends on the temperature, pH, pollutant concentration, contact time, and the amount of the adsorbent used. The magnetic carbonized neem seed demonstrated the highest adsorption capacity for the uptake of RB. Based on the information obtained on the adsorption capacities of the tested isotherms, the magnetic carbonized neem seed (MCNS) showed the greatest affinity towards RB dye, followed by the carbonized form, and the least was observed with the raw neem seed adsorbent. This study, therefore, revealed that both the heat and chemical activation of the raw neem seed improve the textural properties of the material thereby enhancing the adsorption potency of the adsorbent which offers suitable adsorbents for the remediation wastewater.

References

1. Foo, K.Y. and Hameed, B.H., Utilization of biodiesel waste as a renewable resource for activated carbon: Application to environmental problems. *Renew. Sust. Energ. Rev.*, 13, 9, 2495–2504, 2009a.
2. Foo, K.Y. and Hameed, B.H., Value-added utilization of oil palm ash: Superior recycling of the industrial agricultural waste. *J. Hazard. Mater.*, 172, 523–531, 2009b.
3. Al-Rashed, S.M. and Al-Gaid, A.A., Kinetic and thermodynamic studies on the adsorption behavior of rhodamine b dye on duolite C-20 resin. *J. Saudi Chem. Soc.*, 16, 209–215, 2012.
4. Rajae, G., Omar, S., Rachid, T., Kinetic and thermodynamic approaches on Rhodamine B adsorption onto pomegranate peel. *Case Stud. Chem. Environ. Eng.*, 3, 100078, 2020.
5. Jinbei, Y., Shuyue, Y., Wentao, C., Yibing, C., Rhodamine b removal from aqueous solution by CT269DR resin: Static and dynamic study. *Adsorpt. Sci. Technol.*, 37, 9–10, 709–728, 2019.
6. Xiao-Cheng, D., Ji-Hua, Z., Zheng-Jun, Q., Xi-Cun, W., Adsorption of rhodamine b by organic porous materials rich in nitrogen, oxygen, and sulfur heteroatoms. *New J. Chem.*, 45, 3448, 2021.
7. Ani, I., Hadi, N., Mardi, S., Djoko, H., Adsorption study of rhodamine b and methylene blue dyes with ZSM-5 directly synthesized from bangka kaolin without organic template. *Indones. J. Chem.*, 20, 1, 130–140, 2020.

8. Nhamo, C., Edna, C.M., Willis, G., Synthesis, characterization and methyl orange adsorption capacity of ferric oxide–biochar nano composites derived from pulp and paper sludge. *Appl. Water Sci.*, 2016.
9. Santhi, T., Prasad, A.L., Manonmani, S., A comparative study of microwave and chemically treated *acacia nilotica* leaf as an eco friendly adsorbent for the removal of rhodamine b dye from aqueous solution. *Arab. J. Chem.*, 7, 4, 494–503, 2014.
10. Thi, H.Y.D., Thi, P.M.C., Thi, D.D., Thi, H.N., Thi, C.T.V., Nhat, M.N., Bao, H.N., The An, N., Tien, D.P., Adsorptive removal of rhodamine b using novel adsorbent-based surfactant-modified alpha alumina nanoparticles. *J. Anal. Methods Chem.*, 37, 3, 268–289, 2020.
11. Zainab, M.S., Various adsorbents for removal of rhodamine b dye: A review. *Indones. J. Chem.*, 21, 4, 1039–1056, 2021.
12. Said, A., Aly, A., El-Wahab, M., Soliman, S., El-Hafez, A., Helmey, V., Goda, M., Application of modified bagasse as a biosorbent for reactive dyes removal from industrial wastewater. *J. Water Resource Prot.*, 5, 10–17, 2013.
13. Ofudje, E.A., Sodiya, E.F., Ibadin, F.H., Ogundiran, A.A., Alayande, S.O., Osideko, O.A., Mechanism of Cu^{2+} and reactive yellow 145 dye adsorption onto eggshell waste as low-cost adsorbent. *Chem. Ecol.*, 37, 3, 268–289, 2020.
14. Ofudje, E.A., Awotula, A.O., Hambate, G.V., Akinwunmi, F., Alayande, S.O., Olukanni, O.D., Acid activation of groundnut husk for copper adsorption: kinetics and equilibrium studies. *Desalin. Water Treat.*, 86, 240–251, 2017.
15. Mittal, H. and Mishra, S.B., Gum ghatti and Fe_3O_4 magnetic nanoparticles based nanocomposites for the effective adsorption of rhodamine b. *Carbohydr. Polym.*, 101, 1255–1264, 2014.
16. El Qada, E.N., Allen, S.J., Walker, G.M., Influence of preparation conditions on the characteristics of activated carbons produced in laboratory and pilot scale systems. *Chem. Eng. J.*, 142, 1–13, 2008.
17. Ofudje, E.A., Adeogun, I.A., Idowu, M.A., Kareem, S.O., Ndukwe, N.A., Simultaneous removals of cadmium (II) ions and reactive yellow 4 dye from aqueous solution by bone meal-derived apatite: Kinetics, equilibrium and thermodynamic evaluations. *J. Anal. Sci. Technol.*, 11, 1, 1–16, 2020.
18. Oyetade, O.A., Nyamori, V.O., Martincigh, B.S., Jonnalagadda, S.B., Effectiveness of carbon nanotube–cobalt ferrite nanocomposites for the adsorption of rhodamine B from aqueous solutions. *RSC Adv.*, 5, 22724, 2015.
19. Bychko, I., Kalishyn, Y., Strizhak, P., TPR study of core-shell Fe@Fe3O4 nanoparticles supported on activated carbon and carbon nanotubes. *Adv. Mater. Phys. Chem.*, 2, 17–22, 2012.
20. Chen, W.H., Ye, S.C., Sheen, H.K., Hydrothermal carbonization of sugarcane bagasse via wet torrefaction in association with microwave heating. *Bioresour. Technol.*, 118, 195–203, 2012.
21. Yu, Y., Liu, M., Yang, J., Characteristics of vanadium adsorption on and desorption from humic acid. *Chem. Ecol.*, 34, 548–564, 2018.

22. Ofudje, E.A., Akiode, O.K., Oladipo, G.O., Adedapo, A.E., Adebayo, L.O., Awotula, A.O., Application of raw and alkaline-modified coconut shaft as a biosorbent for Pb^{2+} removal. *Bioresources*, 10, 2, 3462–3480, 2015.
23. Akeremale, O.K. and Olaseni, S.E., Comparative studies on the adsorption of rhodamine b and malachite green from simulated wastewater onto bentonite clay. *ChemSearch J.*, 10, 2, 30–40, 2019.
24. Kooh, M.R.R., Dahri, M.K., Lim, L.B.M., The removal of rhodamine b dye from aqueous solution using *casuarina equisetifolia* needles as adsorbent. *Cogent Environ. Sci.*, 21140553, 2016, 2016.
25. Weber, W.J. and Morris, J.C., Kinetic of adsorption on carbon from solution. *J. Sanit. Eng. Div. Proc. Am. Soc. Civ. Eng.*, 89, 31–60, 1963.
26. Akinhanmi, T.F., Ofudje, E.A., Adeogun, A., II, Aina, P., Joseph, I.M., Orange peel as lowcost adsorbent in the elimination of Cd (II) ion: Kinetics, isotherm, thermodynamic and optimization evaluations. *Bioresour. Bioprocess.*, 7, 34, 2020. https://doi.org/10.1186/s40643-020-00320-y.
27. Lagergren, S., Zur theorie der sogenannten adsorption geloster stoffe. *Kungl. Svenska Vetenskapsakad. Handl.*, 24, 1–39, 1898.
28. Ho, Y.S. and McKay, G., Kinetic models for the sorption of dye from aqueous solution by wood. *J. Environ. Sci. Health Part B Process Saf. Environ. Prot.*, 76, 4, 183–191, 1998.
29. Langmuir, I., The adsorption of gases on plane surfaces of glass, mica, and platinum. *J. Am. Chem. Soc.*, 40, 1361–1403, 1918.
30. Freundlich, H.M.F., Über die adsorption in lösungen. *Z. Phys. Chem.*, 57, 385–470, 1906.
31. Ofudje, E.A., Williams, O.D., Asogwa, K.K., Awotula, A.O., Assessment of Langmuir, Freundlich and Rubunin-Radushhkevich adsorption isotherms in the study of the biosorption of Mn (II) ions from aqueous solution by untreated and acid-treated corn shaft. *Int. J. Sci. Eng. Res.*, 4, 7, 1628–1634, 2013.
32. Adeogun, A.I., Ofudje, E.A., Idowu, M.A., Ahmed, S.A., Biosorption of Cr (VI) ion from aqueous solution by maize husk: Isothermal, kinetic and thermodynamic study. *J. Chem. Soc. Pak.*, 34, 6, 1388–1396, 2012.
33. Neto, J.F.D., Pereira, I.D.S., da Silva, V.C., Ferreira, H.C., de A Neves, G., Menezes, R.R., Study of equilibrium and kinetic adsorption of rhodamine b onto purified bentonite clays. *Ceramica*, 64, 598–607, 2018. http://dx.doi.org/10.1590/0366-69132018643722429.
34. Hossain, M.A. and Alam, M.S., Adsorption kinetics of rhodamine-b on used black tea leaves. *Iranian J. Environ. Health Sci. Eng.*, 9, 2, 2012. http://www.ijehse.com/content/9/1/2.
35. Zhu, Z., Xiang, M., Li, P., Shan, L., Zhang, P., Surfactant-modified three-dimensional layered double hydroxide for the removal of methyl orange and rhodamine b: Extended investigations in binary dye systems. *J. Solid State Chem.*, 288, 121448, 2020.

36. Wu, J., Yang, J., Huang, G., Xu, C., Lin, B., Hydrothermal carbonization synthesis of cassava slag biochar with excellent adsorption performance for rhodamine b. *J. Clean. Prod.*, 251, 119717, 2020.
37. Khan, M.A., Momina, Siddiqui, M.R., Otero, M., Alshareef, S.A., Rafatullah, M., Removal of rhodamine b from water using a solvent impregnated polymeric dowex 5WX8 resin: Statistical optimization and batch adsorption studies. *Polymers*, 12, 2, 1–12, 2020.
38. Zhang, J., Hu, X., Yan, X., Feng, R., Zhou, M., Xue, J., Enhanced adsorption of rhodamine b by magnetic nitrogen-doped porous carbon prepared from bimetallic ZIFs. *Colloids Surf. A Physicochem. Eng. Asp.*, 575, 10–17, 2019.
39. Zahir, A., Aslam, Z., Aslam, U., Abdullah, A., Ali, R., Bello, M.M., *Paspalum notatum* grass-waste-based adsorbent for rhodamine b removal from polluted water. *Chem. Biochem. Eng. Q.*, 34, 2, 93–104, 2020.
40. Yu, L. and Ya-Juan, L., Review biosorption isotherms, kinetics and thermodynamics. *Sep. Purif. Technol.*, 61, 229–242, 2008.

5

Green Water Treatment for Organic Pollutions: Photocatalytic Degradation Approach

Yahiya Kadaf Manea[1,2*], Amjad Mumtaz Khan[1], Ajaz Ahmad Wani[1], Adel A. M. Saeed[2], Shaif M. Kasim[2], Ashrf Mashrai[3]

[1]*Department of Chemistry, Aligarh Muslim University, Aligarh, India*
[2]*Department of Chemistry, University of Aden, Aden, Yemen*
[3]*Department of Pharmacy, University of Science and Technology, Ibb, Yemen*

Abstract
Ecological sustainability requires further development in water treatment and management of energy resources. Recently, excessive levels of organic pollutants have been detected in wastewater worldwide. This work highlights new advances in green chemistry for wastewater purification, i.e., solar radiation, photolysis, and titanium/tin-based nanocomposite-mediated photocatalysis. This chapter also describes photodecomposition processes with green technology for organic contaminants mostly found in polluted waters, such as dyes, phenols, and antibiotics. Moreover, the combined effect between the nanocomposites and pollutants with a plausible mechanism, kinetics, and mineralization degree is also presented. This study will be valuable for developing novel nanocatalysts for rapid, economical, and high-performance wastewater treatment.

Keywords: Ecological sustainability, organic pollutions, nanocomposites, photocatalytic

**Corresponding author*: kodaf2006@gmail.com

Shahid-ul-Islam, Abid Hussain Shalla and Mohammad Shahadat (eds.) Green Chemistry for Sustainable Water Purification, (107–128) © 2023 Scrivener Publishing LLC

5.1 Introduction

It has been shown that the release of hazardous wastewater into the ecosystem is a substantial cause of pollution, eutrophication, and disturbance of aquatic life [1, 2]. Organic pollution residues have been identified in practically all environmental matrices on every continent during the last three decades. Surface water (seawater, lakes, rivers, and streams), groundwater, wastewater treatment plant effluent, and sludge are all included in this category [3]. Effluents often contain a variety of chemical substances that are mutagenic, carcinogenic, teratogenic, or otherwise hazardous, putting human health at risk [4, 5]. Due to these concerns, there have undoubtedly been tougher rules on wastewater emissions, which must be considered in a variety of polluting operations. Wastewater treatment is often required in numerous situations. Advanced oxidation processes (AOPs) make it possible to oxidize organic chemicals that would normally be difficult to convert into end products that are less toxic to living organisms [6, 7]. Advanced oxidation processes are characterized by forming hydroxyl free radicals (OH), which are very potent and non-selective chemical oxidants [8, 9]. Comparing the hydroxyl radical to other well-known oxidants, such as ozone and hydrogen peroxide, the hydroxyl radical has greater relative oxidation activity. Following the characteristic features of organic substrates, the produced hydroxyl radicals degrade the organic molecules in a variety of ways, such as radical abstraction, radical addition, electron transfer, and radical combination [10, 11]. When a semiconductor material is excited by electromagnetic irradiation of appropriate energy, the electrons in the valence band are excited into the conduction band. The photocatalytic process is one example of an AOP in which a semiconductor material is excited. The production of electron-hole pairs is the result of this process. The excited electrons can generate superoxide radical ions from the dissolved oxygen, which are highly reactive [12]. The holes generated are also powerful oxidizing agents that may oxidize adsorbed water or hydroxyl radicals to create reactive hydroxyl radicals in the presence of oxygen. It is anticipated that these two radical species would destroy aqueous organic contaminants. TiO_2 has been the most extensively utilized photocatalyst material up to this point with many other catalytic materials, also being explored as photocatalysts [13]. The commercial TiO_2 catalyst, which comprises around 80 weight percent anatase and 20 weight percent rutile, is the most often used catalyst in photocatalytic processes and is available in various sizes. Although it is greatly reliant on the kind and concentration of the pollutant being treated, its optimal activity depends heavily on these factors. Doping TiO_2 with transition metals and nonmetals has been investigated for the last decade, and this combination has been found almost ideal for photocatalysis [4]. The photocatalytic method is effective on a wide range of organic substrates, including diverse hazardous organic pollutants [14, 15], volatile organic compounds. Although the photocatalytic process has great promise, its industrial applicability has remained restricted owing to the high operating costs associated with the process [16]. Specifically, the purpose of this chapter is to demonstrate the applicability of the green photocatalysis approach for the treatment of an aquatic environment contaminated with organic contaminants in general. While this chapter focuses on the degradation of organic pollutants, such as dyes, toxic organic compounds, and antibiotics, it also includes studies of the degradation of model compounds in synthetic and real wastewater. The effects of kinetic energy modeling have been employed to determine the specific applied energy (E_{SAE}) required to degrade the model pollutants. Because AOPs are often energy-intensive processes, electric energy accounts for a significant portion of the operational expenses. The suggested E_{SAE} value provides for quick assessment of system expenses and indicators of the overall amount of power required. It may also be used to estimate the costs associated with system operation.

5.2 Solar Energy

Solar light is the most important source of energy for all life on the planet. Sunlight continuously reaches the earth's surface by a continuous flow of electromagnetic radiation waves throughout the day. Only a very tiny fraction of the total emitted energy in the solar system gets absorbed by the earth's atmosphere.

5.3 Green Photocatalysis

Photocatalysis is a form of catalysis that involves various reactions when light is applied to a surface of the photocatalyst. Photocatalysis includes some critical responses, such as photoactivation of catalysts and degradation of organic molecules via different pathways. The characteristic features of an ideal photocatalyst include stability, cost-effectiveness, efficiency, and nonhazardous. Figure 5.1 shows the simplest schematic photo-driven degradation processes of organic pollutants.

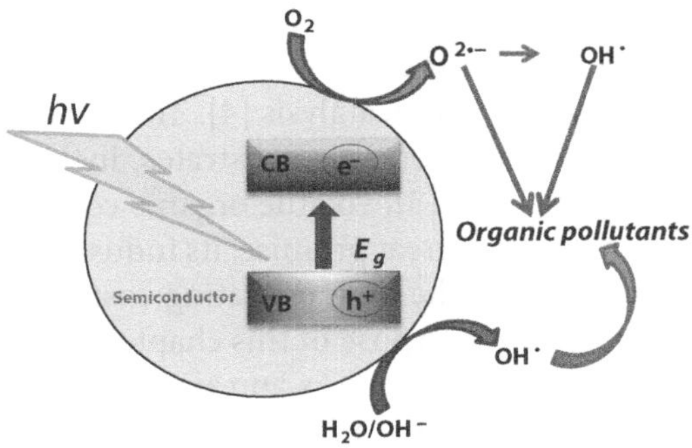

Figure 5.1 Illustration of photodegradation process of organic contaminates.

The semiconductor materials that possess sufficient bandgap could act as sensors for photo redox processes due to their electronic properties. Materials that possess a filled valence band and an empty conduction band absorb light of a specific wavelength, and the electrons present in VB are excited to the CB. The excited electrons to conduction band act as an energy source to be used directly to degrade organic pollutants or create different types of species through attacking O_2 molecules (Figure 5.1). There has been a slew of technologies developed throughout the years to expel toxic contaminants from wastewater. Treatment procedures like adsorption, ion exchange, air, and stripping capture the pollutants but cannot break them down into harmless compounds. The fact that there is no need for after-treatment or disposal procedures makes the photocatalytic process one of the most attractive options compared to other technologies. Another benefit of this procedure is that it does not need costly oxidizing chemicals since ambient oxygen serves as the oxidant. An optimum catalyst should possess the necessary bandgap, absorb a broad range of the solar spectrum, dissociate water molecules, destroy organic contaminants, and must be stable in an aqueous environment during the reaction process.

5.4 Organic Pollutants

Specifically, four distinct model compounds were chosen for this research to reflect the various kinds of contaminants found in wastewaters [17]. Pharmaceutical contaminants are still unregulated, and their residues in the environment are classified as "compounds of emerging concern" because they have a significant effect on the ecosystem and human health. PCOU, also known as 3-(4 hydroxyphenyl)-2 propenoic acid (PCOU), is an organic chemical often discovered in agroindustrial wastes like olive pressing and wine distillery wastewaters. PCOU is prevalent in southern Europe, particularly in the Mediterranean area [6]. It is classified as a contaminant because it interferes with standard biological wastewater treatment. Agriculture wastewaters are thought to include phenolic compounds, which have been linked to increased toxicity and antibacterial activity. As a result, removing these chemicals is critical.

2-Benzofuran-1,3-diene 9 (phthalic anhydride [PHA]) is a chemical compound discharged into the environment by different chemical industries. The majority of the emissions originate during the synthesis of PHA, plasticizers, insecticides, and surfactants, all of which require phthalic acid ester as a component. Besides that, PHA is employed to manufacture a range of products, including polyesters, alkyl resin and polyurethane, halogenated anhydrides, insect repellents, phthalocyanine pigments, dyes, and perfumes, as well as intermediates in the chemical industry [18]. Process off-gases and industrial effluents are the two most significant sources of these emissions. Exposure to phthalic anhydride results in the malfunctioning of the respiratory system irritates the skin and eyes in human beings.

Bisphenol A (2,2-bisphenol A) is a chemical compound found in plants (4-hydroxyphenyl). Phosphoric acid (BPA) is an essential organic chemical that is utilized to synthesize a variety of resins (polycarbonate and epoxy resin type) and a stabilizer during the polymerization process. Compressed discs, dental fillings, plastics, and the support of electrical equipment are all made with BPA. It is possible for BPA to leach into the environment, for example, into surface waters, during production operations, and via migration from finished goods. Because of its estrogenicity, BPA is extensively utilized despite being a recognized endocrine-disrupting chemical (EDC). The EDCs have physiological effects by acting as the natural hormone or interfering with the generation, release, metabolism, and removal of the hormone.

5.5 Reactive Species Responsible for Green Photocatalysis Treatment

Before developing a successful photocatalytic system, it is essential to investigate the function of reactive species during the photocatalytic process. The generation of highly reactive intermediates (holes (h^+), electrons (e), hydroxyl (•OH), and superoxide ion ($O_2^{\bullet}$) radicals) is quite a necessary

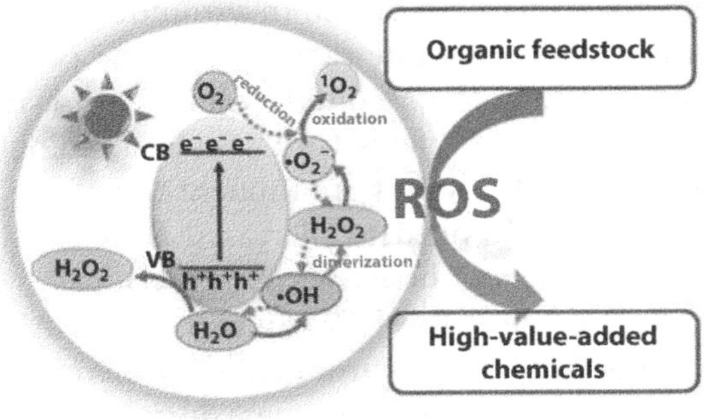

Figure 5.2 Schematic diagram explaining the role of reactive species during the photocatalytic process [19].

step during the photocatalytic reactions. The photochemical reactions carried out by light are the fundamental principle of the semiconductor photocatalytic reactions. During the photocatalytic irradiation process, a light of a specific wavelength falls on the surface of the semiconductor photocatalyst, as a result of this radiation reactive intermediates are produced in the aqueous solution. Organic pollutants are attacked and destroyed by these photogenerated reactive species. During this process, holes are generated in the VB, and electrons are excited to the CB. These photogenerated holes and electrons react with dissolved oxygen and water molecules to produce highly reactive species in different ways, as described in Figure 5.2. Water oxidation produces ·OH and H_2O_2, and dimerization of ·OH can also produce H_2O_2. H_2O_2 can then be oxidized to $^{\cdot}O_2^{-}$, and $^{\cdot}O_2$ and reacts with photogenerated holes to produce $^1O_2$. The O_2 is produced $^{\cdot}O_2$. A proton-coupled process converts O_2 to H_2O_2. Then, H_2O_2 is broken down into ·OH, and $^{\cdot}OH$ is further reduced into H_2O.

5.6 Advancements in Photocatalysts

The development of new nanohybrid as photocatalysts with excellent catalytic and regenerative capacities increases attention in research.

5.6.1 Titanium/Tin-Based Nanocomposite-Mediated Photocatalysis

Various semiconductor materials have been extensively employed as photocatalysts to remediate organic contaminants in recent decades, with the most common being germanium. On the other hand, pure photocatalysts are often hampered in their photocatalytic activities by poor adsorption efficiency, and quick recombination of photogenerated electron-hole pairs, among other factors. Furthermore, their inability to be separated from treated water makes them unsuitable for use in practical applications. Various size-dependent features of semiconductor nanocomposites (SC NCs) have been observed, including electronic bandgap energies, solid-state phase transition temperatures, melting temperature, and compressive behavior. The additional energy needed to excite an electron into the neighboring higher conduction band (CB) in semiconductors allows the electron to flow freely and generate an electric current. This results in a hole being created in the valence band (VB), which travels in the reverse direction of the electron. As a result of their ultra-low bandgap energy, high surface-to-volume ratio, enhanced active sites, self-supporting architecture, and outstanding mechanical qualities, semiconductor nanoparticles are considered as one of the most featured photocatalysts currently available.

The most commonly used photocatalysts for the degradation of toxic pollutants from water and wastewater are Ti-based photocatalysts such as TiO_2, titanium dioxide compounds. Ti is considered as one of the standard reference photocatalysts to compare photocatalytic efficiency under various treatment procedures. A visible light active photocatalyst is extremely desirable, and the amount of light that can be absorbed in appropriate energy depends on the bandgap of the semiconductors. A new form of potential visible light-responsive photocatalyst for organic pollutants, on the other hand, is being developed using tin-based catalysts. Because of its suitable band gap and enhanced chemical stability, it is resistant to degradation. Their photo activities may be increased by modifications in the size and morphology of the catalytic surfaces, which results in a decrease in the recombination of e^- h^+ couples on the catalytic surfaces. A beneficial influence on the photocatalytic action is provided by doping metals or their oxides, which allows charge carrier movement between the photocatalyst and an extra phase, which may delay the recombination of charge carriers. The characteristics of photocatalysts may alter when the interstitial sites are occupied by Ti/Tin and form aggregates on the surface of the catalyst. As a result, the synthesis and use of titanium and tin photocatalysts have piqued the attention of many researchers.

Recently, numerous efforts have been focused on enhancing the physicochemical properties of Tin(IV) and Ti(IV)-based semiconductors, which are highly efficient, stable, and less toxic to our environment as compared to aerogels composed of metal oxides, graphene oxide, and carbon nanotubes. Most of the research work is focused on achieving a high degree of exfoliation, excellent dispersion of (Tin(IV) and Ti(IV))-NPs, and better encapsulation of the active ingredients into the polymer matrices through surface modifications. However, the improvement of such properties is limited to the above influences and is determined by many other factors, including the size, shape, and aspect ratio of the various components. Relevant modeling work has confirmed the importance of these factors on the properties of hybrid materials. Further, it is still a matter of debate how the nanosize affects these mechanical properties due to the limited aspect ratios of the various components. The catalysts' visible light activity has been increased by modifying semiconductors using various techniques, including doping, composite synthesis, size reduction, surface modification, and other techniques. Modification of powdered photocatalysts has been investigated extensively during the past two decades. Table 5.1 shows different kinds of semiconductors photocatalyst and their photocatalytic degradation efficiencies toward different organic pollutants. For example, Ti(IV) phosphate/Poly o-anisidine (TP-POA), SM(III) doped POA-TP, TiO_2@polyaniline, $BiVO_4/TiO_2$, Tin(IV) silicomolybedate@ Acrylamide(TSM/AC), and TSM@Polyacrylamide-g-Chitosan (TSM@ PA-CS) [20, 21]. The following characteristics of the improved photocatalyst are observed:

1. Improved harvesting of visible light.
2. Decrease in the bandgap energy.
3. Catalytic characteristics that have been improved.
4. Increased reusability.

5.6.2 Synthesis of Various Nanocomposites as Photocatalysts

Manea and colleagues have developed a photocatalyst-based titanium (IV) nanocomposite for the breakdown of MB and MG dyes that is very efficient. Poly o-inside @titanium phosphate (POA-TP) nanocomposite catalyst with acidic surface properties was synthesized using the sol-gel process and had a mesoporous structure. To synthesize the POA polymer with inorganic complex TP, an *in situ* oxidative polymerization procedure was carried out in the laboratory. Typically, the gel of POA was added to the white precipitate of TP. At the same time, the combination was constantly stirred, and the resulting mixture was progressively transformed into black-colored slurries over some time. After that, the material was allowed to rest at room temperature for 24 hours. The POA-TP products were filtered out and extensively washed with DMW to eliminate any

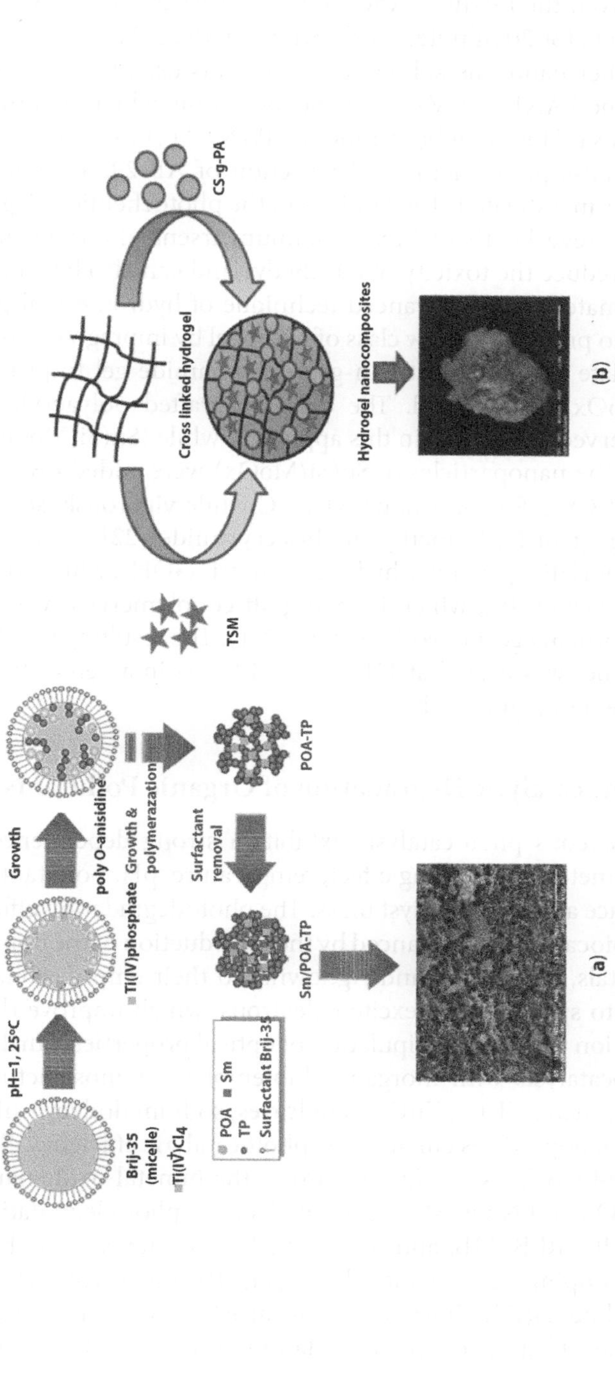

Figure 5.3 (a, b) Schematic diagram showing the route synthesis of (a) Sm@POA-TP and (b) CS-g-PA@TSM nanocomposites [9].

excess acid from the mixture. After multiple washes, the gel was dried in an oven at 40°C for 20 minutes, as shown in Figure 5.3a.

On the other hand, the sol-gel technique was used to create a unique semicrystalline PANI-Ti (IV) as a nanocomposite, which was then characterized. PANI-Ti is an abbreviation for PANI-Ti (IV). Additionally, the capacity to cause photochemical destruction of AB-29, an organic contaminant, was investigated. The findings of the photochemical degradation investigations revealed that (PANI@Titanium arsenate) may be utilized to successfully reduce the toxicity of AB-29 dye and can also be employed as conducting material. The advanced technique of hydrothermal processes was applied to prepare the new class of hydrogel by impregnations Tin(IV) silicomolybdate NPs into chitosan-g-polyacrylamide gel to produce CS/PA-Sn(Si/MoOx) nanohybrid. The chitosan-grafted polyacrylamide gel (CS-g-PA) served as the host in this approach, while (Sn(Si/MoOx) served as the filler. The nanoparticles of Sn(Si/MoOx) were added dropwise into the matrix of CS-g-PA for 3 hours at 70°C while vigorously stirring. The cross-linking agent N, N methylene bisacrylamide [22]. As a result, the redox initiator (KPS) generates hydroxyl radicals (•OH), which form active free-radical sites on PA, which is then graft copolymerized with CS biopolymer to form 3D gel network (Scheme 2-1). The resulting CS/PA-Sn(Si/MoOx) product was heated at 120°C for 24 hours in a Teflon-lined autoclave, as shown in Figure 5.3b.

5.6.3 Photocatalytic Degradation of Organic Pollutants

The heterogeneous photocatalysis exhibits a strong dependence on the process parameters, i.e. doping effect, temperature, pH, pollutant concentration, surface area, and catalyst mass. The photodegradation efficiency of catalysts or biocatalysts is enhanced by the introduction of the doping effect of noble metals, such as Au and Ag, owing to their empty orbital, which could allow to spread of the excited electrons, which improve the visible light absorption with the manipulation of optical properties. Further, combined photocatalysts with inorganic, biotemplates, nanostructures filler, and micelles, such as TiO_2, ZnO, spent leaves, sodium dodecyl sulfate, and metal oxide nanoparticles enhance the photocatalytic efficiency. For example, the use of Co_3O_4 NPs, NiO nanorods, the bimetal oxide nanohybrid of CeO_2@Y_2O_3, and NiO@MnO exhibited strong photodegradation activity toward MO, RhB, MB, and rose bengal dyes. Manea *et al.* have been studying the doping effect by introducing Sm(III) rare earth metal ion into poly-o anisidine-Ti(IV) phosphate to produce Sm/POA-TiP nanocomposite, which was effective for the degradation of MB (94%) and MG (90%)

Table 5.1 Different kinds of semiconductors photocatalyst and their efficiencies for photodegradation of some organic pollutants.

S. no.	Sn(IV) NCs	Synthesis method	Parameters			Potential applications	Efficiency	Ref.
			Sn (Amt)	T(°C)/time	pH			
1	SnO_2-PBS	Sol-gel	Sn^{4+} (0.315 g and 30%)	100°C/4 h	11-12	Photocatalytic degradation of RhB	87%	[23]
2	r-GO –SnO_2	Hydrothermal	Sn^{4+} (0.1M and 30%)	180°C/4 h	9	Photocatalytic degradation of MO	100%	[24]
3	TiO_2@SnS_2	Hydrothermal	Sn^{4+} (0.1M and 25%)	40 °C/24 h	—	K_2CrO_7 reduction	90%	[25]
4	CS/PA@Tin(IV) Si/Mo	Hydrothermal	Sn (0.1M and 25%)	120°C/24 h	1-2	photodegradation CIP	95.5%	[26]
5	AC-Tin(IV) Si/Mo	Sol-gel	Sn^{4+} (0.1M and 25%)	Refluxed 60°C/ 24 h	0.63	MB degradation	90%	[21]

dyes. The most important factors influencing the photocatalytic process are summarized in Table 5.1.

Several researchers have studied the photodegradation of organic pollutants using different semiconductor materials, i.e., transition metal ions and their metal oxides, due to their efficiencies as catalysts for organic pollutant degradation. This chapter also focuses on the recent developments of the mechanism and applications of some nanocomposites based, Tin (Sn^{+4}) and Titanium (Ti^{+4}) as promising candidates. In addition, this article presents a systematic series of studies on preparation with controllable size and shape using a wide range of methods. Relevant growth mechanisms are also discussed. Furthermore, the promising potential of Ti/Sn (IV) NCs as building blocks for energy, sensors, and catalysis is highlighted. Tailor-made nanocomposites combining rare earth metals or conducting polymers are being developed to improve the performance of photocatalyst composites. This chapter can provide researchers with key snapshots of recent advances and future challenges, resulting in significant progress in 2D CNT and biopolymers, as well as M(IV) - semiconductor NCs.

5.7 Green Treatment of Pollutants

The green treatment of antibiotics, dyes and other hazardous compounds by photocatalysis approach is the major chemical process of decomposition. Toxic properties must be reduced and their transport into water systems must be restricted. Table 5.2 shows the most important factors influencing the photocatalytic process.

5.7.1 Photodegradation of Toxic Dyes

The presence of organic dyes in bodies of water, even at low concentrations, is highly undesirable. In general, all dyes cause a slew of health problems in humans, including allergic dermatitis, skin problems, and cancer. Photocatalytic process using highly efficient catalyst/semiconductor is known as green and active approach for treating the toxic dyes. Increasing the absorption capacity toward the solar/visible light area by inducing the photoexcited dye molecule decreases the absorption capacity toward the visible light region. Many dyes, such as methylene blue rhodamine B, porphyrins, malachite green thionine, rose Bengal, erythrosine B, and others, can harvest visible light, which is why they are utilized as sensitizers. When the energy level of the dye molecule is more negative than the energy level of the semiconductor, electrons are moved from the dye molecule to the

Table 5.2 The most important factors influencing the photocatalytic process.

Factors	Effects
Pollutant concentration	Inorganic pollutant elimination on the surface of the photocatalyst is directly proportional to the initial concentration of organic pollutants. Even if the breakdown percentage of organic pollutants falls with an increase in pollutant concentration, it is necessary to maintain the appropriate amount of photocatalyst in the system.
pH solution	Depending on the nature of the material and the qualities of the pollutant, the pH solution may either stimulate or prevent photocatalytic activity.
Temperature-dependent reaction	To obtain effective photocatalytic activity, the reaction temperature should be in the range of 0–80°C. If the temperature is higher than 80°C, the catalyst will enhance the recombination of the electron-hole pair and reduce the system's photocatalytic activity.
Surface area	The use of nanomaterials with large surface areas may improve the photocatalytic performance by increasing the number of active sites (AS) on the photocatalyst surface, resulting in the formation of more radical reactive species, which is necessary for effective photodegradation activity.
Doping on the catalyst surface	There are several ways available for intercalating active ions, which can capture photons with very little energy. Bandgap engineering is achieved by doping metals and nonmetals into photocatalytic materials, which may then be used to adjust and continually alter valence and conduction bands in the photocatalytic materials.

semiconductor's conduction band that they are transported. The processes involved, photoexcitation, electron injection, and dye regeneration, as explained in Figure 5.4.

For example, Cinelli and his colleagues has been investigated the degradation efficiency of RhB^+ dye using $CoFe_2O_4$/BiOCl composite. The experimental results demonstrated that visible light photocatalysis induced by $CoFe_2O_4$/BiOCl for Rhodamine B degradation was successful, and spectroscopic analysis revealed the presence of a hypochromic shift 554 nm to 495 nm (Figure 5.4). Furthermore, it has been discovered that photobleaching

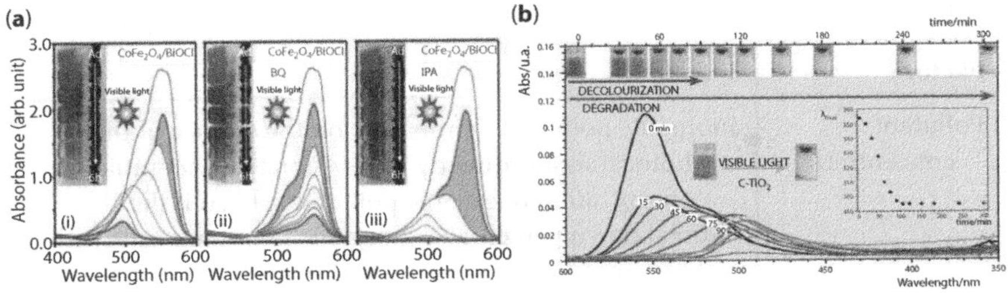

Figure 5.4 (a) Photodecomposition of RhB dye over $CoFe_2O_4$/BiOCl composite under visible irradiation (i) without and with scavengers of (ii) BQ and (iii) IPA [27]. (b) UV-vis spectrum for decolorization and photocatalytic degradation of Rhodamine B [28].

is required for the production of RhB^+, whereas percent OH is responsible for the N-demethylation reaction. Because of RhB's oxidation potential and TiO_2's band edges, the excited dye may inject electrons into the TiO_2 conduction band, resulting in particles that become cationic radicals and undergo a further transition to products. It is possible that the photosensitization response includes the following reactions.

Furthermore, it has been discovered that photobleaching is required for the production of RhB^+, whereas percent OH is responsible for the N-demethylation reaction. Because of RhB's oxidation potential and TiO_2's band edges, the excited dye may inject electrons into the TiO_2 conduction band, resulting in particles that become cationic radicals and undergo a further transition to products. It is possible that the photosensitization response reaction was suggested as following:

$$RhB + hv \longrightarrow RhB^* + TiO_2 \longrightarrow RhB^+ + TiO_2\,(e)$$

5.7.2 Photodegradation of Antibiotics

The presence of emerging contaminants (ECs) in hospital sewage, such as pharmaceutical compounds [anti-inflammatory drugs (NSAIDs), steroidal hormones, and antibiotics (e.g., ciprofloxacin, tetracycline, Amoxicillin, salbutamol, estrone, benzocaine, and others)] [28–30], has been observed in wastewater around the world. So, it is essential to utilize an effective method to cleanse wastewater that contains these contaminants before it can be put to any good use. According to the researchers, improvement in the photodegradation of antibiotics with the help of nanocatalysts may be ascribed to their better visible light absorption capacity and their ability to separate photogenerated electron-hole pairs efficiently. Furthermore, the huge surface area of the photocatalyst has the potential to increase the effectiveness of photocatalytic degradation. The active sites on the catalyst surface may be responsible for forming free radicals (during photocatalysis) on the catalyst surface. After being grabbed (and transferred to O_2 molecules), the electrons were adsorbed on the surface. They generated reactive species (radical dot OH, e^-, O^{2-}, and h^+) released into the environment. As a result of the above facts, we previously described a correct photocatalytic degradation process of ciprofloxacin utilizing a tin-based nanocomposite (CS/PA-Sn(Si/MoOx)), which is depicted by the schematic diagram in Figure 5.5.

$$\text{Catalyst} + h\nu \rightarrow \text{Catalyst}^* + e^- + h^+$$

$$h^+ + H_2O \rightarrow .OH + H^+$$

$$e^- + O_2 \rightarrow O^{2-}$$

$$^{\bullet}O^{2-} + {}^{\bullet}OH + H^+ \rightarrow H_2O_2 + {}^{\bullet}O_2$$

$$O^{2-} + {}^{\bullet}OH + h^+ + \text{Ciprofloxacin} \rightarrow CO_2 + H_2O + \text{other anions}$$

5.7.3 Photodegradation of Bisphenol BPA

In water pipes, medical equipment, dental sealants, epoxy resins (such as plastic water bottles and polycarbonate plastics), numerous food containers, thermo bond, electronics, and toys, the chemical bisphenol (BPA) may be found in high concentrations. Because it resembles the human hormone estrogen, the chemical is harmful to the reproductive system. Inorganic metabolism, such as organism growth, reproduction, metabolic systems, brain networks, and cardiovascular irrigation, are negatively affected by bisphenol A (BPA) in the environment. It has been discovered that BPA may be found in almost every environmental compartment investigated, including the air, water, and soil [31]. According to published research, BPA exposure is associated with elevated anxiety, sadness, hyperactivity, and inattention. There is evidence of bisphenol in the organic body, which suggests that it is present in the blood and urine and a risk factor for cardiovascular disease, diabetes, and obesity. It also appears to be a risk factor for fetal development and reduces basal testosterone secretion in males. Another finding is that BPA and other related compounds have been found

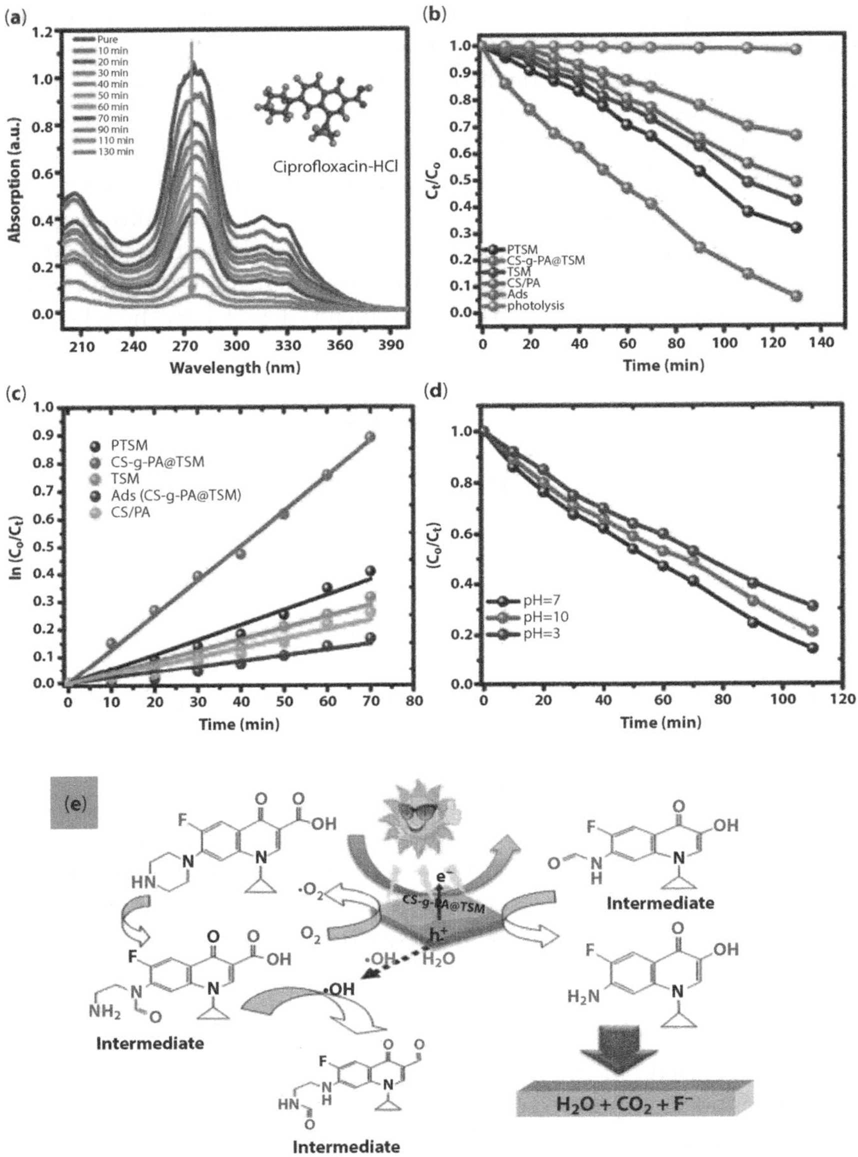

Figure 5.5 Schematically shows photodegradation of antibiotic ciprofloxacin.

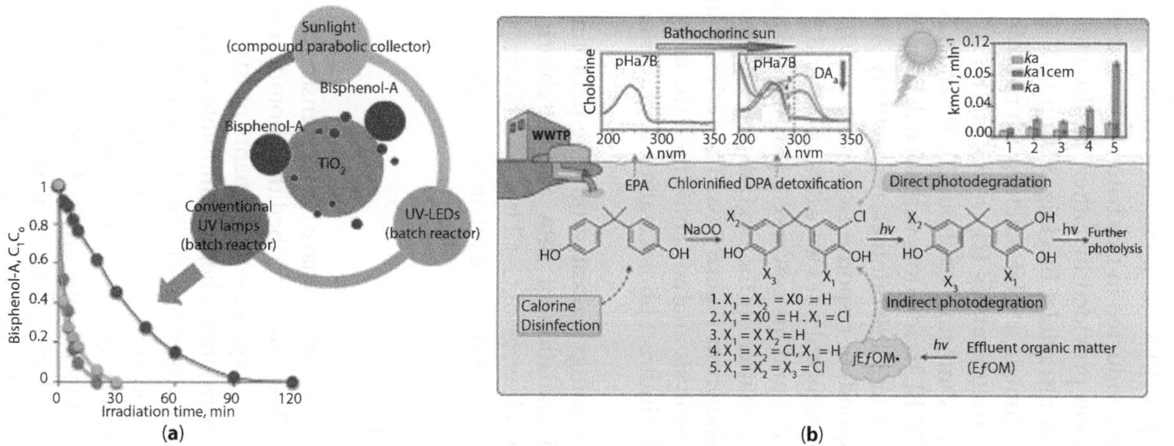

Figure 5.6 (a) Bisphenol-A removal by TiO_2 photocatalyst in three different systems [32]; (b) photodegradation of chlorinated derivatives of BPAs under direct and indirect sunlight [33].

in various environments, foods, and food-holding containers, as well as in human milk, urine, and placental tissue; this is evidence of probable worldwide exposure to the chemicals. David *et al.* investigated the photocatalytic treatment of bisphenol-A (BPA) under various irradiation conditions [32]. The experiments were conducted in two batch-operated slurry photoreactors with UV irradiation provided by either an ultraviolet light-emitting diode (UV-LED) or an ultraviolet blacklight lamp (UV-BL); additionally, the experiments were done in a solar reactor with natural sunlight.

Under optimized circumstances (C_0 = 2.5 mg/L, TiO_2 = 250 mg/L), BPA was destroyed in a short time of about 20, 30, and 120 minutes under UV-LED, solar, and UV-BL irradiation, with total decomposition occurring in approximately 20, 30, and 120 minutes, respectively. After 90 minutes of treatment, the corresponding response rates were 0.230, 0.151, and 0.025 minutes, and the TOC removal rates were 88%, 67%, and 33%, respectively. On the other hand, Wan and his coworkers [33] were explained the effect of the effluent organic matter on the degradation of BPAs and its chlorinated derivatives as shown in Figure 5.6a. The series of quenching studies and laser flash photolysis analysis confirmed the association of triplet states of (EfOM) to the indirect photocatalytic degradation of BPAs with a rate constant $10^9\,M{\cdot}s^{-1}$.

The experimental findings indicated that the photoproducts for direct and indirect photodegradation of BPA and its chlorinated derivatives mainly ascribed to the cleavage of C–Cl bond and hydroxylation with further cleavage of the benzene ring as explained in Figure 5.6b.

5.8 Conclusion

In recent decades, various methods have been extensively used to remediate organic contaminants. We have described the photodegradation approach as a green technology for organic contaminants mostly found in polluted waters, such as dyes, hormones, phenols, antibiotics. This study indicated that the exfoliation and dispersion of (Tin(IV) and Ti(IV))-NPs as active ingredients into the polymer matrices through surface modifications enhance the photocatalytic efficacy. However, the improvement of such properties is limited to the abovementioned influences and is determined by many other factors, including the size, shape, and aspect ratio of the various components. Relevant modeling work has confirmed the importance of these factors on the properties of hybrid materials. Therefore, this chapter will provide valuable information for developing novel nanocatalysts for rapid, economical, and high-performance wastewater treatment.

References

1. Wani, A.A., Khan, A.M., Manea, Y.K., Salem, M.A.S., Shahadat, M., Selective adsorption and ultrafast fluorescent detection of Cr(VI) in wastewater using neodymium doped polyaniline supported layered double hydroxide nanocomposite. *J. Hazard. Mater.*, 416, 125754, 2021.
2. Shahadat, M. *et al.*, Heavy metals scavenging using multidentate/multifunctional aerogels and their composites, in: *Advances in Aerogel Composites for Environmental Remediation*, pp. 275–296, 2021.
3. Patel, M. *et al.*, Pharmaceuticals of emerging concern in aquatic systems: Chemistry, occurrence, effects, and removal methods, 119, 6, 2019, 3510–3673, 2019.
4. Shahadat, M., Teng, T.T., Rafatullah, M., Arshad, M., Titanium-based nanocomposite materials: A review of recent advances and perspectives. *Colloids Surf. B Biointerfaces*, 126, 121–137, 2015.
5. Manea, Y.K. *et al.*, In-grown flower like Al-Li/Th-LDH@CNT nanocomposite for enhanced photocatalytic degradation of MG dye and selective adsorption of Cr (VI). *J. Environ. Chem. Eng.*, 106848, 2021.
6. Pirilä, M. *et al.*, Photocatalytic degradation of organic pollutants in wastewater. *Top. Catal.*, 58, 1085–1099, 2015.
7. Minh, D.P., Gallezot, P., Azabou, S., Sayadi, S., Besson, M., Catalytic wet air oxidation of olive oil mill effluents: 4. Treatment and detoxification of real effluents. *Appl. Catal. B Environ.*, 84, 749–757, 2008.
8. Sultana, S., Rafiuddin, Khan, M.Z., Shahadat, M., Development of ZnO and ZrO2 nanoparticles: Their photocatalytic and bactericidal activity. *J. Environ. Chem. Eng.*, 3, 886–891, 2015.
9. Manea, Y.K., Khan, A.M., Nabi, S.A., Facile synthesis of mesoporous sm@ POA/TP and POA/TP nanocomposites with excellent performance for the photocatalytic degradation of MB and MG dyes. *J. Alloys Compd.*, 791, 1046–1062, 2019.
10. Mantzavinos, D., Hellenbrand, R., Livingston, A.G., Metcalfe, I.S., Catalytic wet oxidation of p-coumaric acid: Partial oxidation intermediates, reaction pathways and catalyst leaching. *Appl. Catal. B Environ.*, 7, 379–396, 1996.
11. Baransi, K., Dubowski, Y., Sabbah, I., Synergetic effect between photocatalytic degradation and adsorption processes on the removal of phenolic compounds from olive mill wastewater. *Water Res.*, 46, 789–798, 2012.
12. Qahtan, A.A.A., Husain, S., Somvanshi, A., Khan, W., Manea, Y.K., Influence of Mn doping on dielectric properties, conduction mechanism and photocatalytic nature of gadolinium-based orthochromites. *J. Mater. Sci. Mater. Electron.*, 31, 12, 9335–9351, 2020.
13. Teh, T., Nik Norulaini, N.A.R., Shahadat, M., Wong, Y., Mohd Omar, A.K., Risk assessment of metal contamination in soil and groundwater in Asia: A review of recent trends as well as existing environmental laws and regulations. *Pedosphere*, 26, 431–450, 2016.

14. Abdullah, A., Abdullah, A.Z., Ahmed, M., Okoye, P.U., Shahadat, M., A review on bi/multifunctional catalytic oxydehydration of bioglycerol to acrylic acid: Catalyst type, kinetics, and reaction mechanism. *Can. J. Chem. Eng.*, 1–30, 2021.
15. Wani, A.A., Khan, A., Manea, Y.K., Salem, M.A.S., Enhanced photocatalytic degradation of organic dyes from aqueous environment using neodymium-doped mesoporous layered double hydroxide. *J. Rare Earths*, 2021.
16. Manea, Y.K. and Khan, A., Membrane technology for water treatment: Design, development, and applications, in: *Chemistry and Industrial Techniques for Chemical Engineers*, Apple Academic Press Inc., Taylor & Francis Group, Canada, pp. 1–33, 2019.
17. Wani, A.A. *et al.*, Graphene-supported organic-inorganic layered double hydroxides and their environmental applications: A review. *J. Clean. Prod.*, 273, 122980, 2020.
18. Shahadat, M. *et al.*, Clay-based adsorbents for the analysis of dye pollutants, in: *Applied Water Science Volume 1*, 163–197, 2021.
19. Luo, L., Zhang, T., Wang, M., Yun, R., Xiang, X., Recent advances in heterogeneous photo-driven oxidation of organic molecules by reactive oxygen species. *ChemSusChem*, 13, 5173–5184, 2020.
20. Manea, Y.K., Khan, A.M., Nabi, S.A., Facile synthesis of mesoporous sm@POA/TP and POA/TP nanocomposites with excellent performance for the photocatalytic degradation of MB and MG dyes. *J. Alloys Compd.*, 791, 1046–1062, 2019.
21. Khan, A.M., Manea, Y.K., Nabi, S.A., Synthesis and characterization of nanocomposite acrylamide TIN(IV) silicomolybdate: Photocatalytic activity and chromatographic column separations. *J. Anal. Chem.*, 74, 330–338, 2019.
22. Manea, Y.K. *et al.*, Hydrothermally synthesized mesoporous CS-g-PA@TSM functional nanocomposite for efficient photocatalytic degradation of ciprofloxacin and treatment of metal ions. *J. Mol. Liq.*, 335, 116144, 2021.
23. Kar, A. *et al.*, Facile synthesis of SnO2-PbS nanocomposites with controlled structure for applications in photocatalysis. *Nanoscale*, 8, 2727–2739, 2016.
24. Sagadevan, S. *et al.*, A facile hydrothermal approach for catalytic and optical behavior of tin oxide-graphene (SnO2/G) nanocomposite. *PloS One*, 13, e0202694, 2018.
25. Li, J., Sun, W., Dai, W.M., Zhang, Y.C., Hydrothermal synthesis and photocatalytic properties of TiO2/SnS2 nanocomposite. *Adv. Mater. Res.*, 898, 23–26, 2014.
26. Manea, Y.K., Khan, A.M.T., Wani, A.A., Qashqoosh, M.T.A., Shahadat, M., Mansour, A.S.S., Hydrothermally synthesized mesoporous CS-g-PA@TSM functional nanocomposite for efficient photocatalytic degradation of ciprofloxaci9n and treatment of metal ions. *J. Mol. Liq.*, 335, 116144, 2021.

27. Cinelli, G. *et al.*, Photocatalytic degradation of a model textile dye using Carbon-doped titanium dioxide and visible light. *J. Water Process Eng.*, 20, 71–77, 2017.
28. Qashqoosh, M., Alahdal, F., Manea, Y.K., Synthesis, characterization and spectroscopic studies of surfactant loaded antiulcer drug into Chitosan nanoparticles for interaction with bovine serum albumin. *Chem. Phys.*, 527, 110462, 2019.
29. Manea, Y.K., Khan, A.M.T., Qashqoosh, M.T.A., Wani, A.A., Shahadat, M., Ciprofloxacin-supported chitosan/polyphosphate nanocomposite to bind bovine serum albumin: Its application in drug delivery. *J. Mol. Liq.*, 292, 111337, 2019.
30. Manea, Y.K. *et al.*, Interaction of AMOT@CS NPs and AMOT drug with bovine serum albumin: Insights from spectroscopic and molecular docking techniques. *Chem. Phys.*, 546, 111139, 2021.
31. Sin, J.C., Lam, S.M., Mohamed, A.R., Lee, K.T., Degrading endocrine disrupting chemicals from wastewater by TiO 2 photocatalysis: A review. *Int. J. Photoenergy*, 2012, 1–23, 2012.
32. Davididou, K. *et al.*, Photocatalytic degradation of bisphenol-a under UV-LED, blacklight and solar irradiation. *J. Clean. Prod.*, 203, 13–21, 2018.
33. Wan, D. *et al.*, Formation and enhanced photodegradation of chlorinated derivatives of bisphenol a in wastewater treatment plant effluent. *Water Res.*, 184, 116002, 2020.

6

Treatment of Textile-Wastewater Using Green Technologies

Shuchita Tomar[1], Mohammad Shahadat[2*], S. Wazed Ali[1], Mangala Joshi[1] and B.S. Butola[1]

[1]Department of Textile and Fibre Engineering, Indian Institute of Technology Delhi, Hauz Khaz, New Delhi, India
[2]School of Chemical Sciences, University Sains Malaysia, Penang, Malaysia

Abstract

The textile industry has become the largest water consumption industry in the world. Wastewater released from textile industries is comprised of several pollutants, like synthetic dyes, cleaners, steadying agents, chemicals, inorganic salts, and heavy metals. Many textile industries are generally responsible for releasing unprocessed wastewater into rivers, lakes, etc., thus causing serious health threats to humans, aquatic life, plants, etc. Textile wet processing methods generate effluents like chemical oxygen demand (COD), biological oxygen demand (BOD), reactive dyes, chemicals, and various organic compounds. Therefore, the textile effluents need to be treated before being released into rivers, lakes, etc. Elimination of pollutants from water bodies includes various methods, such as physicochemical, biological, combined treatment methods, and other different techniques. Researchers are still doing research to process textile wastewater in a more sustainable, cost-effective, and effective way. Nowadays, sustainable green wastewater treatment methods are gaining more importance in the textile industry. It is necessary to implement an eco-friendly green technology method for wastewater treatment to avoid adverse effects on the environment and ecological balance. In this chapter, we mainly discuss the different green technologies for treating textile wastewater.

***Keywords*:** Population growth, green technology, textile dye pollution, removal of dye, human health

**Corresponding author*: mdshahadat93@gmail.com

Shahid-ul-Islam, Abid Hussain Shalla and Mohammad Shahadat (eds.) Green Chemistry for Sustainable Water Purification, (129–156) © 2023 Scrivener Publishing LLC

6.1 Introduction

Textile industries contribute a lot in maintaining the economy of countries worldwide. The textile industry itself contributes almost 12-14% of total production to the market economy. According to a survey, in 2020 the market value of the textile industry has been increased to around 4% from 2019 and reached 842.6 billion dollars. With the rise in textile products demand, the number of textile industries increases and also concurrently increase in wastewater pollution, thus causing severe health hazards to the people. Generally, in textile industries, the yarn spinning method such as ring spinning, winding, etc and weaving of fabric on the loom is a kind of dry process and does not consume that much water but in wet processing methods like singeing, desizing, scouring, bleaching, mercerizing, dyeing, printing, finishing, etc, consumes maximum amount of water and thus discharging various chemicals like dyes, acids, alkalis, dispersing agents, suspended solids in large amount, and hence, generating polluted wastewater. In a study, it was found that, commercially, around 100,000 types of dyes would be there and at a worldwide level, approximately 1,000,000 tons of dyes were produced annually. Also, it was stated that each year textile dyes of nearly 280,000 tons were released as industrial waste in water bodies all over the world. Therefore, it was observed that most of the dyes are extremely toxic, hazardous, reduce diffusion of light and photosynthetic activity that can cause a shortage of oxygen and restrain downstream valuable uses like drinking water, regeneration, and irrigation. Hence, causing serious health problems to humans like cholera, jaundice, etc. Thus, it is very essential to treat textile wastewater by eliminating the pollutants from water. Earlier, conventional wastewater treatments technology become responsible for rising the landfill and wide-ranging infrastructure needs, and also it affects the ecosystem as the landfill management is not done correctly. Moreover, this technology requires extensive energy that subsequently can emit carbon. Some of the textile wastewater treatments are done using biological processes followed by chemical coagulation. Though chemical and biological processes are best suited for eliminating dyes, they need a particular technique and are energy-intensive; furthermore, a high level of by-products is produced that generates a safe discarding issue. Nowadays, green sustainable technologies are mainly focused to treat wastewater for an eco-friendly environment. Sustainable green technology wastewater treatments give substantial pollutant removal efficiency, and significantly decrease the management of slush, consumption of energy, and running cost [1, 2].

This process mostly involves biologically built treatments providing greater organic exclusion efficacy and reducing the adversative effect on the ecosystem, thus providing sludge management through a sustainable way that devours less energy and decreases the inclusive cost [3–6]. This chapter mainly focuses on environmental challenges in textile industries and the importance of green technology for wastewater treatment and different methods for treating textile effluents [7, 8].

6.1.1 Textile Industries: Causes of Water Pollution

The wastewater generated by textile industries usually depends on the coloration of textiles and chemicals used in different stages of processing and manufacturing. The dyeing of textiles could be done by using several kinds of dyes like acid dye, direct dye, reactive dye, vat dye, disperse dye, etc. Mostly dyes are made up of organic and inorganic complexes that can be applied in various ways to textiles. Generally, exhaustion of dyes ranges from 50% to 85% and thereby released as a pollutant in effluents due to which water bodies got affected severely. The discharged wastewater of colorants comprises around 4000 to 5000 parts per million (ppm) of suspended solids. However, the wastewater released by textile industries contains high levels of BOD, COD, total solids, chemicals, salts, residual dyes, acid/alkalis, some cleaning solvents, etc that can cause a serious threat to human health as well as water species [9, 10]. The different processing stages in textile manufacturing that are mainly responsible for generating pollutants are shown in Figure 6.1. Initially, the textile fibers are converted into yarns

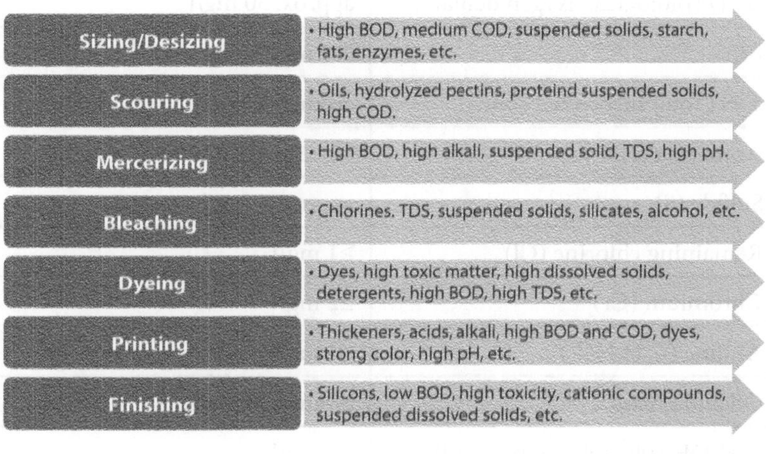

Figure 6.1 Wastewater characteristic for various levels of the textile processing industry.

through spinning process then they are finally woven into fabrics and run further for various wet processing stages (dyeing, finishing, etc). Wet processing generally requires a high amount of water during processing and thus discharges wastewater in rivers, lakes, etc. Various wet processing methods are more responsible for producing pollutants and effluents in large quantities (sizing, scouring, bleaching, mercerizing, dyeing, printing, finishing). However, among them, the dying process produces more wastewater. Around 15% of textile dyestuff was utilized in the manufacturing of products that got released worldwide annually into the environment. Usually, reactive dyes are widely used for dyeing textiles because of having extensive diversity of shades of color, high wet fastness property, easy application, intense colors, and nominal energy requirement. Thus, from textile dyeing industries reactive dyes are generally responsible for contributing high levels of wastewater in effluents. Moreover, some of the dyes, such as aromatic and heterocyclic dyes, show complex and firm structures so they might become more hazardous in degradation when existing in textile wastewaters. In addition to this spinning process also produce wastewater. The melt and the dry spinning process do not produce that much wastewater, but the wet spinning process produces a large amount of effluent,

Table 6.1 Standards for effluent released by textile processing industries in India.

Constraints	Concentration apart from (pH)
COD (Chemical oxygen demand)	approx. 250 mg/l
BOD (Biological oxygen demand)	approx. 30 mg/l
pH	6-9 mg/l range
TSS (Total suspended solids)	around 100 mg/l
Ammonia (NH_3)	12-15 mg/l
Sulfide (S)	approx. 1 mg/l
Remaining chlorine (Cl)	≥1 mg/l
Chromium (Cr)	≥2 mg/l
Aniline ($C_6H_5NH_2$)	≥1 mg/l
Oil and grease	around 10 mg/l
C_6H_5OH (phenolic compound)	≥1 mg/l

Table 6.2 Standards for domestic effluent discharge [11].

Parameters (mg/l) except pH	Domestic surface water	Municipal sewers	Domestic irrigation	Seaside areas
pH	5.5 to 9.0	5.5 to 9.0	5.5 to 9.0	5.5 to 9.0
Total suspended solids	1.0×10^2	6.0×10^2	2.0×10^2	1.0×10^2
Total dissolved solids	21.0×10^2	-	21.0×10^2	-
Biochemical oxygen demand	3.0×10^1	3.50×10^2	1.0×10^2	1.0×10^2
Chemical oxygen demand	2.50×10^2	-	-	2.50×10^2
Sulfate (SO_4^{2-})	10.0×10^2	10.0×10^2	10.0×10^2	-
Chloride (cl^-)	10.0×10^2	10.0×10^2	6.0×10^2	-
Oil or grease	1.0×10^1	2.0×10^1	1.0×10^1	2.0×10^1
Lead	1.0×10^{-1}	1.0	-	2.0

such as the wet spinning of viscose rayon generates a high number of pollutants in wastewater. Table 6.1 shows the number of pollutants generated by wet and dry spinning operations in textile industries. The textile industry in India set some specific values for effluent discharge that are listed in Table 6.1. Moreover, some specific values are also set for domestic effluent discharge that is listed in Table 6.2.

6.1.2 The Effect of Polluted Water Discharged From Textile Industries on the Environment

Textile industries generally consume a large volume of water along with some chemicals during different steps of processing and at last releases wastewater containing pollutants and various coloring agents in rivers, lakes, etc. In the dyeing process, the uptake and dyes fixation degree on the substrate is not proper thus remaining dyes discharge with wastewater where some dyes are water-soluble, and some are water insoluble. However, it was assumed that around 2% of dyes were released directly into effluents, and around 10% vanished in the coloration process. Thus, it was expected that almost 20% of colorants were released into the environment through textile wastewater and hence causing a serious environmental hazard to human health and aquatic life. Treatment of textile wastewater is very tough because of having various chemical compositions, high color intensity, etc. Some of the dyes persist for a longer duration in the environment. Among them, azo (dye) under aerobic conditions is usually resilient to biodegradation while in the anaerobic condition they release aromatic amines as the azo bond shows easy reductive splitting. Effluent released from dye industries includes high organic content and high COD and BOD values. Wastewater released from dye and pigment industries contains insoluble organic pigments, dyes, emulsifiers, and heavy metals as shown in Table 6.3. The BOD and COD values do not give complete information about the toxicity level of wastewater. Thus, it can be achieved through acute and enrichment toxicity tests [12]. Generally, the acute/chronic effect of dyes is based on dye concentration and exposure time on organisms. The greatest environmental distress is sunlight absorption and reflection in water having dyes. Absorption of sunlight may disturb the activity of photosynthesis of algae and thus deploy the food chain. In addition to it, degradation of dyes via bacteria is known and it can be précised by several parameters like oxygen, temperature, pH, dye concentration, dye structure, carbon concentration, and other bases of nitrogen. Therefore, to achieve efficient dye degradation, it is a must to observe the consequence of these parameters on the biodegradation of dyes. Table 6.4 shows some aspects that affect the degradation of dyes. Few dyes and their by-products used in textile industries show carcinogenic, mutagenic behavior and can cause liver, kidney cancer in humans. Some textile dyes can also cause skin irritation,

Table 6.3 Dyes and pigment waste released by textile industries.

Colorants	Properties
Diazotization-based coupling reaction, the reaction by condensation process, nitration, sulfonation, reduction, and oxidation.	High organic content, high BOD (2×10^2 to 4×10^2 mg/l) and high COD (5×10^2 to 2×10^3 mg/l) and their ration up to 0.4, high TDS, high acidic pH, Co^{+2}, Cr^{+6}, Cu, etc.
Tint	
Grinding using ball mills, washings	Strong colors, high BOD and COD, organic matter turbidity.

Table 6.4 The parameters affecting the bacterial degradation of dyes.

Parameters	Details
Effect of pH	For eliminating colors, the range of pH for bacterial biodegradation is 6 to 10 [13].
Effect of Temperature	In the case of azo dyes rate of decolorization rises to an ideal temperature and then there is a peripheral decrease in decolorization.
Concentration of dyes	In the case of dye concentration, as dye concentration increases toxicity will also increase and thus decrease in bacterial decolorization rate.
Carbon, nitrogen bases	Dye-biodegradation needs carbon and nitrogen-based organic sources (extract of yeast, carbohydrates peptone, or organic compounds) so that degradation and decolorization can occur.
Structure of dye	The low molecular weight structure of Dyes shows better decolorization.
Oxygen	It was observed that in oxidative enzymes activity some quantity of oxygen is needed for better degradation of azo dyes.

respiratory diseases, allergenic reaction to eyes, dermatitis, etc. Hence, for the safety of human health, the elimination of textile dyes color from water is necessary after that other soluble colorless organic matter and other pollutants from water need to be removed.

6.1.3 Various Techniques for Effluent Treatment

Wastewater discharged from textile industries contains toxic dyes, chemicals, and other auxiliaries in which maximum of them are added while doing textile processing. Dye colors exist in textile wastewater usually inhibiting the activity of photosynthesis and also affecting the growth of aquatic species thus disturbing the ecological balance. Moreover, contaminated water can cause various diseases to humans, such as diarrhea, skin irritation, nausea, ulceration of skin, etc. Thus, water needs to be treated for removing these types of toxic compounds and colors to avoid any harm to humans and ecological activity. Various types of methods are adopted to eliminate the contamination from wastewaters such as physical methods,

Wastewater treatment techniques		
Primary treatment 1. Screening 2. Sedimentations 3. Homozenization 4. Neutralization 5. Mechanical flocculation	**Secondary treatment** 1. Aerobic and anaerobic method 2. Activated sludge method 3. Trickling filteration 4. Oxidation	**Tertiary treatment** 1. Membrane method 2. Adsorption 3. Oxidation 4. Electrochemical coagulation 5. Ion exchange 6. Thermal evaporation

Figure 6.2 The techniques for treating wastewater of textile industries 2.

chemical methods, biological methods, and eco-friendly methods or green treatment methods. Figure 6.2 shows that the effluent treatment process occurred in different sections, primary treatment, secondary treatment, and tertiary treatment. Within this, primary treatment usually eliminates suspended solids, floating and determined substances, after this secondary treatment helps in oxygen demand reduction, chemicals, and pollutant colors and finally, the tertiary treatment part eliminates all the remaining contaminations from the pollutant. After the treatment process water quality should be measured before using it or discharging into rivers, lakes, etc. [14–17].

6.1.4 Physical Treatment Technique

This technique helps to eliminate the pollutants from wastewater through various forces like electric magnetism, gravity, Van der Waals force, and by any means of physical barrier. This technique does not change the material structure chemically, but any physical change or coagulation of materials that exist in water can happen. Few techniques have been conferred here.

6.1.4.1 Adsorption Method

The adsorption method is one of the most useful physicochemical methods used for treating wastewater. It shows higher efficiency in eliminating pollutants as compared to other methods. Generally activated carbon, silicon polymers, and kaolin adsorbents show better dye adsorption ability. Ions or molecules in this method normally exist in the gas or liquid phase and got accumulated on the solid-phase matter. Physical adsorption occurs when there is feeble interspecies attraction among adsorbate and adsorbents but if there is robust interspecies attraction among them then because of electrons exchange chemical adsorption take place. Moreover, the process is contingent on the surface area of the adsorbent, size of the particle, contact time, the concentration of adsorbent and adsorbate, pH, and temperature. Activated carbon is the most widely used adsorbent for effluent treatment. This helps in adsorbing metal ions, many dyes, cations, and mordants efficiently. For sustainable green treatment methods, they can be manufactured from a carbonaceous substance like sawdust, coconut husks, materials made from cellulosic, palm kernel husks, peanut skins and hulls, coffee, tea waste, modified sugar beet pulp, maize leaves, etc. and give better adsorption. Further in this chapter, the physical bioadsorption method is also discussed in detail [2, 11].

6.1.4.2 Ion-Exchange Method

This method is used to purify, separate, and eliminate contaminations of ion and aqueous solutions. This technique is not much preferred for wastewater treatment as it cannot treat all types of dyes. They are only able to eliminate unwanted cationic or basic dyes, anionic or acid, direct and reactive dyes effectively. Some common ion exchangers are zeolites, resins, montmorillonite, clay, soil humus, etc. few amphoteric ion exchangers are capable to change cations and anions both concurrently and also used effectively in mixing beds comprising of cation and anion exchange resins mixture. Some of the important benefits of this method are adsorbent retrieval, solvent reclamation after the use of solvent, and efficient elimination of soluble dyes [18].

6.1.4.3 Floatation

In the floatation method, air at high pressure is provided under which suspended organic compounds or particles get attached with bubbles of air that are less dense as compared to water and thus make a group of suspended particles, which helps in increasing the surface area of feed wastewater. There are different kinds of flotation methods used for treating effluents like dissolved air flotation, precipitate flotation, and ion flotation. But this process has some disadvantages, such as energy consumption is high, operational cost, including intense mechanical operations is also high. Hence, it is best suitable for maintenance workplace areas [19].

6.1.5 Chemical Treatment Technique

To remove pollutants from wastewater chemicals are used so that water can be purified and used again. Chemical procedures used show chemical reactions and along with this chemical unit processing physical and biological processes are also applied. The various type of chemical methods used for wastewater treatment are chemical precipitation, coagulation, and flocculation, chemical oxidation, Fenton oxidation, etc. [2].

6.1.5.1 Chemical Precipitation Method

This is the most widely used chemical method for removing pollutants and other toxic materials from wastewater. In this method, the metals that exist in wastewater in a dissolved state are changed into solid particles because of the addition of a precipitate reagent. Precipitate reagent initiates the chemical reaction to take place that converts the dissolved metals to solid form and thus eliminated by the filtration process. The type of precipitation method to be used usually depends on the type of metals that exist, their concentration level, and also the type of reagents used. For example, in the case of the hydroxide precipitation method, sodium or calcium hydroxides are mostly required as a reagent to change dissolved metals into solid particles, although this is quite tough to form hydroxides as wastewater comprises mixed metals [2].

6.1.5.2 Coagulation and Sedimentation Method

This method has been widely used earlier. In this procedure, a few chemicals are mixed with water and then charged particles form compounds that coagulate in water. Generally, the coagulants are organic or inorganic having a positive charge in water. Many organic polymers create a good level of coagulation and hence coagulants produce sediments that can be extracted easily. Some of the chemicals are used frequently, such as $FeCI_3$, $AI_2(SO_4)_3$, $FeSO_4$, lime, etc. [20].

6.1.6 Chemical Oxidation

In the chemical oxidation technique, some of the chemicals are used to eliminate the pollutants from wastewater either by demolishing them or by neutralizing their harsh effects. This process can also be utilized along with the physical treatment process to get better results. Chemical oxidation treatment of textile wastewater can be done by oxidizing the pigment present in dyeing and printing or by bleaching it. Among all the different chemical methods, Fenton oxidation and ozone oxidation are used widely for treating wastewater. Photocatalytic oxidation and sonocatalytic oxidation methods are also useful in this technique. O_3 and H_2O_2 chemicals are generally used in this method that makes robust hydroxyl radicals at a high pH, which efficiently breaks conjugated double bonds of chromophore dye group and complex aromatic rings thus, decreasing the wastewater color effectively [15].

6.1.6.1 *Ozonation Method*

The ozonation technique is the utmost efficient technique that breaks the double bond of dye molecules and helps in removing the color of textile wastewater. It helps in oxidizing a significant quantity of COD and restrains the foaming behavior of remaining surfactants. Moreover, it also helps in increasing the biodegradability rate of the effluent by converting the high amount of nonbiodegradable and toxic substances into biodegradable mediates. One of the benefits of this method is that there is no sludge production. Ozone is a good oxidizing agent as it shows high reactivity and efficiently degrades phenols, chlorinated hydrocarbons, aromatic hydrocarbons, and pesticides. The major disadvantage of this method is that ozone has a short life span because it got decays within 20 minutes, hence this technique requires a constant supply of ozone that is very costly [22].

6.1.6.2 *Fenton Oxidation Method*

In this method, hydrogen peroxide is generally used as an oxidizing agent that makes hydroxyl radicals which helps in decolorizing various dyes from textile wastewater. In the Fenton reaction, hydroxyl radicals are formed and got activated when an acidic solution is added in H_2O_2 having Fe^{2+} ions. In large-scale plants, at higher temperatures reactions are done and also a high amount of iron and H_2O_2 are utilized and a high amount of COD is eliminated. One of the disadvantages of this method is the high level of sludge production and elevated effluent release.

6.1.6.3 *Evaporation*

Evaporation is one of the significant methods under tertiary wastewater treatment techniques. It is used to dry up the solutions that are concentrated such as retrieval of caustic from mercerized and separation of solid substances that are dissolved, dyes, water through dye bath, and the excluded water from both nanofiltration and RO process also be separated via evaporation by doing pure water vaporization and condensation [63]. In textile effluent treatment, RO saturates and condensed water from the evaporator gives pure distilled water that is again utilized in different textile processing or boilers. Two evaporation methods are generally used for treating textile effluent: solar evaporation and mechanical evaporation. [23–25].

Figure 6.3 Solar-evaporation-based plant grafted in textile industries.

Table 6.5 Advantages and disadvantages of solar evaporation method.

Advantages	Disadvantages
Construction is simple	The rate of evaporation is very low.
Less energy cost required	Large surface area is required.
Very less maintenance required	In the rainy season more chances of contamination to be overflowed into groundwater.
No need for highly skilled workers	Water vapor goes into the atmosphere causing fewer chances for retrieval of pure water for recycling purposes.

6.1.6.4 *Solar Evaporation Method*

The saline effluent goes into the large surface area within open tanks, which is known as solar ponds having 0.5 to 0.75 mt height. Various solar ponds that are constructed get filled up regularly and need to be operated and maintained properly on regular basis. Surface water gets evaporated during sunlight and vapors are gone into the atmosphere as shown in Figure 6.3. Therefore, all the dissolved solids or the dehydrated substance in concentrated form which settled down at the bottom got eliminated frequently. But there are some advantages and disadvantages of this treatment technique as given below in Table 6.5.

6.1.7 Mechanical Evaporation Method

In this method, mechanical-based evaporators are provided heat through steam condenses on metallic and cylindrical tubes. When steam got condenses on the outer side of the tube, condensation heat makes the saline water slurry to become boiled. This technique is useful in desalination, dehydration for dilute concentrated solutions, separate pure water from effluents. In textile industries, for treating saline effluents multiple-effect evaporation plants (MEEP) have been used. In this, first, dye bath then washes followed by RO or nanorejects after that fluid crystallizer used to separate water and solid matter. Batch or continuous operating evaporators can be used, and their configuration may be vertical or horizontal. MEEP, a vertical cell carrying a stainless-steel tube, which is heated using steam and the concentrated liquid part got collected at the bottom of the evaporator after that it was sent to next treatment level of the evaporator, where at the first level of evaporator freshwater has been added. In the meantime, steam can also be provided to the first level of evaporator so that the next level got steam from the preceding one. The liquid is circulated again and again

Table 6.6 Multiple effect evaporation plant (MEEP) operating parameters [28].

Required properties	Level I	Level II	Level III	Level IV
Temp (°C)	83±2	72±2	62±2	55±2
Cylindrical Tube distance (mt.)	6.0	6.0	4.5	4.5
Cylindrical Tube dia (mt.)	90.0	90.0	45.0	45.0

till the required concentration is reached at the time of evaporation. The parameters of MEEP are given below in Table 6.6 [21, 26, 27].

6.2 Green Water Treatment Technique for Textile Effluents

The development of green water treatment techniques has gained wide interest because of having various benefits to our ecological balance. Green technology is also helpful in preserving non-renewable energy sources and reducing the misuse of water. Moreover, this technique reduces pollution and mishandling of natural resources. Nowadays, this technique is also become important to keep the biodiversity, their habitats, and especially aquatic species life safe. Here, some of the sustainable and green water treatment techniques were discussed briefly.

6.2.1 Electrocoagulation (EC)

Electrocoagulation is one of the sustainable techniques for treating textile industry wastewater. This technique is generally based on electrochemical methods that eliminate the suspended solids, colloidal matter, metals, also some other dissolved solids present in wastewater. In this process, constituents are broken down by passing electricity which is known as "electrolysis," basically, this process involves direct current (DC) supply to metal electrodes due to which reaction occur on electrode water interface and convert from neutral to a charged state. Thus, charged metal ions enters the water by leaving the plate, and hence, electricity is transferred to electrodes and form charged metal coagulants that make bond along with pollutants as the passed electricity weakens the suspended solids and precipitate out from the solution. The main benefit of this technique is that it produces less amount of sludge as compared to other coagulation methods like the chemical methods. Here, dewatering of sludge is easier that substantially decreases the cost of disposal. This technique can treat wastewater by eliminating heavy metals, suspended solids, bacteria, pesticides, etc. present in water. Its eliminating efficacy is more than 95% and can be a better pretreatment to membrane method in which a large quantity of water can be again utilized. Table 6.7 shows the electrocoagulation treatment of some of the different types of dyes present in textile wastewater [29–31].

Table 6.7 The electrocoagulation process of various dyes present in textile wastewater.

Dye	Density of current	Anode & cathode	Dye removing efficacy	References
Orange-84 (reactive dye)	130 $A.m^{-2}$	SS & SS, Fe & Fe	66% & 76%	[32]
Black (B), Orange (3R), Yellow (GR) (reactive dyes)	$6.25 \times 10^{-2} A.cm^{-2}$	Al & Al	Upto 98%	[33]
Black (5) (reactive dye)	0.75×10 $mA. cm^{-2}$	Fe & SS	Upto 90%	[34]
RB 133 (remazol red dye)	15 $mA.cm^{-2}$	Al & Al	90-92%	[35]
Red 23 (direct dye)	30 $A.m^{-2}$	Fe & Fe, Al & Al	>95%	[36]
Red 14 (acid dye)	80 $A.m^{-2}$	Fe & St 304	Up to 93%	[37]
Orange7 (acid dye)	5 $mA.cm^{-2}$	Boron doped diamond & copper foil	Up to 98%	[38]
Red (disperse dye)	20.8 $mA.cm^{-2}$	Al & Al	90-95%	[39]
RR 241 and DB 60	75 and 50 $A.m^{-2}$	Fe & Fe SS & SS, Al & Al-SS & SS	Upto 99%	[40]

6.2.2 Advanced Oxidation Process (AOP)

Nowadays, the advanced oxidation process (AOP) has also been preferred for treating textile effluent. Organic, inorganic substances and pollutants are eliminated by hydroxyl or sulfate radicals discharged in a specific amount and also increase the biodegradability rate of water. AOP type, pollutant physical and chemical characteristics, and operating parameters affect the efficiency of this method. AOP is an important technique for removing contaminants from industrial wastewater as it shows the good oxidative ability of ozone and formed OH radicals. AOP based on ozone has been widely considered in simulated and real conditions. The role of auxiliary agents presents in dye that helps in dye degradation and also the consequence of different salts on ozonation was observed through AOP. This method is also very useful in leachate treatment and recycling of water. Various types of AOP that can be applied are O_3, O_3/H_2O_2, O_3/UV, UV/TiO_2, UV/H_2O_2, Fenton reaction, photo-Fenton based reaction, ultrasonic irradiation, heat or persulfate, UV or persulfate, Fe^{2+} or persulfate, OH/persulfate. Using AOP we can form and use hydroxyl free radicals. One other process based on Activated carbon catalyzed ozonation (ACCO) helped enhance the oxidation of organic compounds and thus inspires stimulated carbon in a radical form chain reaction of ozone decomposition and OH also formed. In the case of the chloride ions effect, an increase in the concentration of NaCl effectively remove dyes like RB194. ACCO performance can also be improved by more chlorine radicals creating acidic nature. Hence, this technique becomes an effective color removal technique from saline textile wastewater [41].

6.2.3 Rotating Biological Contactor (RBC)

This method has been used since the early 1900 for treating domestic wastewater. RBC consists of various sizes of glass containers known as reactors; circular disc arrangement set comprised of a polymeric substance such as polystyrene or PVC. The disc around 40% is emersed in wastewater and the shaft stay above the surface of the water this whole process is assisted by a motor. In this biological treatment method, biofilm-based filters were used in circular disc form, which was fixed to a horizontal shaft that slowly rotates to gather the contamination. This could be immersed partly or entirely in wastewater, and these RBS shafts remain parallel or perpendicular towards the effluent flow. According to some researchers, it was found that when reaction among the microbial community and RBC film takes place, then biofilms behave as an active source of bacteria having high biochemical potential, which is very useful in eliminating the contaminations having phenol compounds from aqueous medium. The main advantage of RBC is a high level of concentration of biomass, which shows high efficacy, and also bacterial study can be done correspondingly. The medium, efficacy of wastewater treatment, bacterial community configuration all are interrelated and help in the adsorption process. RBC process utilizes low energy, is easy to handle, requires lesser area, is usually aerobic, and helps in eliminating biodegradable matter, also in BOD and COD elimination therefore generally utilized in the textile industry. In the case of the biological treatment process, the wastewater and microorganisms get interconnected among the layer of biomass in the RBC disc which shows biodegradation and also decreases the nitrate level in the effluent. Moreover, the disc in an organized order having several stages is more useful in eliminating the maximum quantity of organic matter and as the biofilm got air exposure because of rotation, the biofilm can absorb oxygen present in wastewater. Also, the existing microbial culture allows entire dye adsorption from wastewater. It is necessary to eradicate biomass when it gets exceeded the limit. The main member of this process is the rotating disc that gives interfacial space in between the biofilm and the impurities and helps in removing the pollutants [42].

6.2.4 Sequencing Batch Reactor (SBR)

SBR method shows better biodegradation of both sulfonated azo and diazo reactive dyes from textile effluent and it is also good for removing phosphorous and nitrogen from piggery waste. According to the research studies, adding amines to SBR culture degrades the aerobic bacteria and increases the efficiency of azo dye elimination. In one of the studies, it was stated that novel microbial consortia by using SBR shows a very good elimination of blue Bezaktiv S-GLD 150 dye from wastewater, around 90% to 97% decolorization occur and up to 98% COD eliminated from effluent. Also, it was found that biological treatment method both aerobic and anaerobic along with NF shows good results in treating textile effluent [43].

6.2.5 Effluent Treatment Using Enzymes

Biological catalysts and their actions can cause chemical reactions in which enzymatic systems are involved that also occur among the two chemical and biological processes. Some substances that are harmful to the environment can be grown by the enzymatic method. The enzymatic method is one of the virtuous methods for those that contain a high concentration of contaminants or the lowest concentration of different contaminants which tends to inhibit enzymes' vulnerability to inactive in presence of added chemicals. It was observed that white-rot fungal cultures can decolorize different types of eight synthetic dyes, as well as azo, anthraquinone metal complex, and indigo. Ligninase-based catalyzed oxidation can decolorize these dyes around 80%, and their decolorization rate can be increased by increasing the ligninase dosage. Chrysosporium culture is capable to degrade the entire azo and heterocyclic dyes. *P. tremella, P. ostreatus, B. adusta*, and *C. versicolor* are four types of white-rot fungi that are used as an enzymatic treatment for removing textile dyes. Nowadays, enzyme membrane reactors are developed for treating textile wastewater, especially for the depolarization of dyes [44].

6.2.6 Membrane Filtration

The membrane filtration technique has been developed as a sustainable and economical method that can eliminate dyes from textile wastewater. It can decrease the coloration level, BOD, and COD in textile wastewater. Ultrafiltration is one of the membrane filtration methods in which macro size particles within a range of 1 nm to 0.05 μm got separated but it cannot separate diluted substances or nanosize particles so the treated water cannot be reused. Therefore, the nanofiltration method is much preferred for the removal of dyes from textile effluent. The combination of adsorption and nanofiltration is adopted for effluent treatment in which adsorption occurs first, and then the nanofiltration process is used to reduce the effluent concentration. Membranes of nanofiltration can reserve the divalent ions of low molecular weight, large monovalent ions, hydrolyzed type reactive dyes, dyeing auxiliaries, etc. This method helps in treating the extremely complicated solutions and also environment-friendly method. Table 6.8 shows some research done on the nanofiltration technique for the removal of dyes from textile effluent [2].

6.2.7 Bioadsorbents Process for Effluent Treatment

The main mechanism of the adsorption process is based on interaction among the surface of the cell and positive dye ions. Living biomass surface cell is made up of polysaccharides, proteins, lipids and also contain negative charges that can attract the maximum amount of positive dye ions present in wastewater. Azo, nitro, and hydroxyl groups present in dye molecules can increase the adsorption process. Microbial biomass shows the ability to adsorb because of the mechanism of the exchange of ions.

Table 6.8 Eco-friendly nanofiltration membranes for textile dye removal.

Nanofiltration membranes	Dyes	Removal efficiency	References
PVDF loose nanofiltration (NF) membrane with multilayer structure modified with GO and PPy.	Negative dyes (dye/salt solution)	98.5%	[45]
Chitosan (CS) based NF membrane	Methyl viologen (MV), methylene blue, orange G, rose bengal, brilliant blue G250	Up to 99.9%	[46]
PEI-based membrane	RR120	81%	[47]
Coating AMTHBA and PEI on L-NF membranes	MB AND CR dyes	98-99%	[48]
CA-f-TA nanofiller fabricated on PES-NF membranes	Direct Red 16 dye	95.7%	[49]
CS–EDTA–mGO (M6-NF) membrane	RR195	97.2%	[50]
Helical type carbon incorporated by chitosan L-NF membrane, cross-linked with SDS.	Both cationic and anionic dyes	>90%	[51]

Figure 6.4 shows the dye and cell interaction with each other. There are two types of sorption chemisorption and physisorption and they occur as liquid phase molecules got clutched to the solid phase surface because of the force of attraction in the adsorbent surface, thus disabling the kinetic energy of adsorbate molecules [52].

An adsorption isotherm model is given which shows a curve that relates the concentration of solute on the adsorbent surface (QE) to the

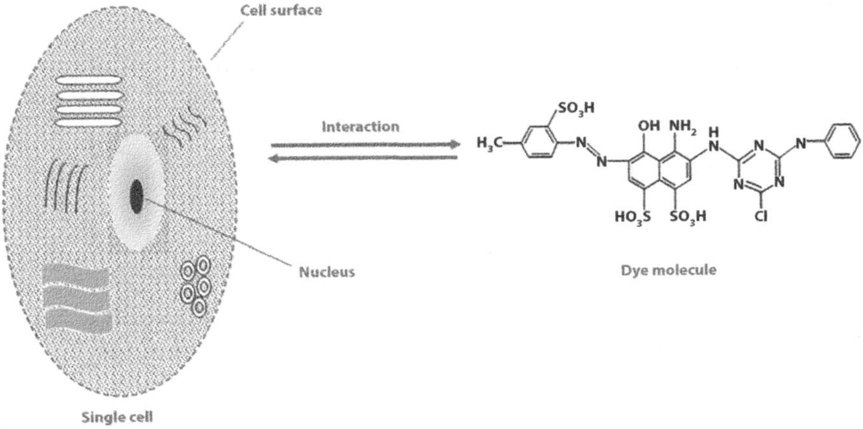

Figure 6.4 The process of interaction between the dye molecule and biomass [10].

concentration of solute in liquid in contact (c_e). The age and Ce parameters relation can be put to more than one isotherm model. There are different equations of isotherm to observe the parameters of sorption equilibrium. In the case of activated carbon adsorption treatment, Freundlich and Langmuir's isotherms are generally used [53].

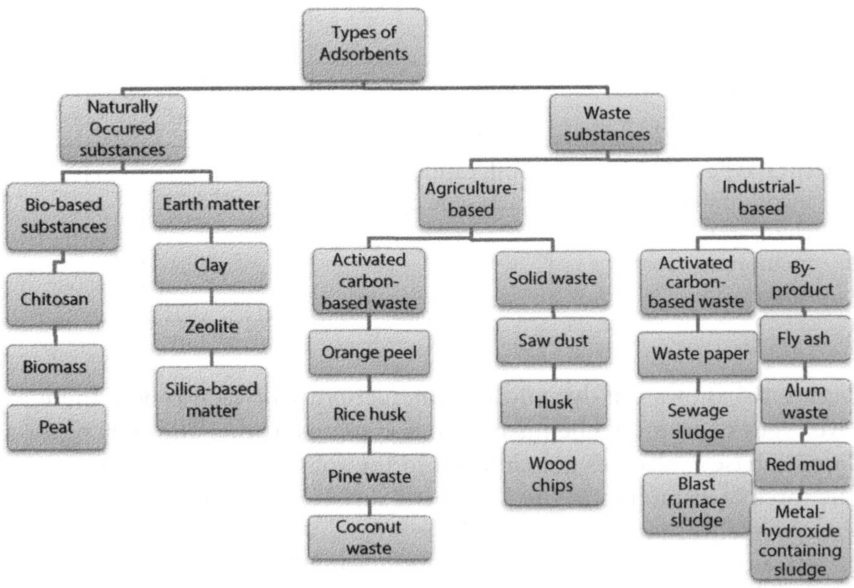

Figure 6.5 The categories of textile-based adsorbents.

Generally, solid-type surfaces show the ability to adsorb as in effluent treatment the efficacy of solid matter depends on its assembly, porosity degree, permeability, specific area, etc. Adsorbate could be organic compound consisting of color, smell, etc. for textile effluent treatment

Table 6.9 Adsorbents used for textile effluent removal.

Bioadsorbent	Dye	Characteristics	References
Wheat flour biodegradable adsorbent based on carbohydrate	Cationic Rhodamine B (RB) dye	Langmuir-isotherm model used and shows high dye adsorption ability.	[55]
Cellulose nanofibers PCFs modified to get (DAMNC) cellulose grafts.	Uranium (VI) metal ion, safranin-o dye, methylene-blue dyes	Langmuir isotherm model used depicts the adsorbate monolayer coverage on DOC.	[56]
Banana waste	Novacron Blue FN-R (reactive dye), methyl red, malachite green dye	Able to change dark colored effluent water to colorless	[57]
Banana, orange peels activated surfaces	Red dye (reactive dye)	Dye reduction. Langmuir and Freundlich isotherm model was used.	[58]
The remainder of waste tea	Blue 25 (acid dye)	Anionic dye bioremediation occur.	[59]
Pd NPs made from cotton boll waste	Toxic azo dye	Effective in degrading the toxicity in waste	[60]
Waste of wheat husk	RB 19 and RB 195	Heavy metal elimination from textile effluent	[61]

adsorbents can be activated carbon, organic polymers, silica-based compounds, etc.

Nowadays, everyone is focusing on the application of organic wastes from agriculture, forest and there is a need to use recycled products in industries. For treating the wastewater of textile industries adsorbents are very helpful in adsorbing the harsh and poisonous substances from wastewater so that water can be reused for different purposes. Figure 6.5 shows the various categories for textile-based adsorbents. This method has gained so much importance as it is eco-friendly green technology, less cost, biodegradable, sustainable, etc. It has been found that utilization of garlic peel waste, sugarcane waste, banana waste, orange waste, banana stalk, coir, and its waste, husk of a coconut, citrus rind, rice hull, and many more organic wastes have shown good results in adsorption of toxic matter from wastewater. They are much better than the costly activated carbon. Some sorbents considerably show very good dye adsorbing ability, such as reactive dyes. This method is simple in use, nontoxic, less cost for effluent treatment. Table 6.9 shows bioadsorbent-based research for removing textile effluent. Some normally applied adsorbents for effluent treatment are given below [54].

6.2.7.1 Citrus Fruits

Citrus fruits are a highly rich source of vitamins and minerals. It has been observed that every year approximately 120 million tons of waste citrus are produced. According to research findings, these types of waste can be utilized for making various productive materials, such as fiber, biodegradable polymer, 3D substances, energy storage materials, etc. Orange peels show good results in adsorbing 1-naphthyl amine dye from wastewater of textile industries.

6.2.7.2 Coir Fiber

Coir fiber is normally used for the removal of color and COD. The layer by layer technique has recently been used for removing pollutants from wastewater. For example, chitosan and polyacrylic when deposited on coir using the layer-by-layer method then their ability to adsorb get enhanced as the energy required during pressure filtration was dazed. It shows good results in removing pollutants from textile wastewater. It has low-cost, easily available, strong, etc. [56, 62].

6.2.7.3 *Coconut Shell–Activated Carbon*

Coconut shell–activated carbon is effective in removing dyes from textile effluent. Sulfuric acid is used for the activation of coconut husk. In a study, it has been reported that maximum blue-GRL and yellow DY-12 (direct dye) can be removed through adsorption. This coconut shell is treated with sulfuric acid that acts as a bioadsorbent. Adsorptions occur through multilayers, which are heterogeneous, endothermic. Methylene blue also shows good result at optimum conditions when 0.1 g/100 ml is used at pH 8, temperature 303°K. Cheaper, easy to use, shows good efficiency for dye removal.

6.3 Conclusions

All over the world, textile industries generate a large number of effluents that pollute the water bodies. Textile effluent comprises a large amount of chemical oxygen demand, biological oxygen demand, TDS, SS, toxic dyes that need to be treated before releasing in water bodies, otherwise it can cause serious diseases to human and aquatic species. Hence, treatment of textile effluent is very necessary to avoid any environmental issues. There are various methods available that can be adopted for treating wastewater like physical methods: adsorption technique, ion exchange, flotation; chemical methods: sedimentation and coagulation, chemical precipitation, chemical oxidation, ozonation; evaporation method, and green technology-based techniques. The key focus in this chapter is to elaborate on various green techniques or sustainable methods for treating textile effluent, such as advanced oxidation process, rotating biological contactor, electrocoagulation, sequencing batch reactor, enzyme treatment, membrane filtration, bioadsorbents. These green techniques are less expensive, easy to use, almost chemical-free, easily available, and show good results in treating textile effluent.

References

1. Singh, M., Chadha, P., Mehra, S., Adverse impact of textile dyes on the aquatic environment as well as on human beings. *Toxicol. Int.*, 28, 2, 165–176, 2021.

2. Mani, S., Chowdhary, P., Bhargava, R.N., Textile wastewater dyes: Toxicity profile and treatment approaches, in: *Emerging and Eco-Friendly Approaches for Waste Management*, pp. 219–244, Springer, Singapore, 2018.
3. Shahadat, M., Bushra, R., Nabi, S., Raeissi, A.S., Development of titanium-supported ion-exchange adsorbent for removal of metal pollutants. *Arab. J. Sci. Eng.*, 43, 7, 3601–3609, 2018.
4. Bushra, R., Shahadat, M., Arshad, M., Preparation of polyaniline based nanocomposite material and their environmental applications. *Int. J. Environ. Sci. Technol.*, 12, 11, 3635–3642, 2015.
5. Nabi, S.A., Raeisi, A.S., Shahadat, M., Bushra, R., Khan, T., Synthesis and characterization of novel cation exchange adsorbent for the treatment of real samples for metal ions. *Chem. Eng. J.*, 200-202, 426–432, 2012.
6. Sk, A.R., Shahadat, M., Basu, S., Shaikh, Z.A., Ali, S.W., Polyaniline/carbon nanotube-graphite modified electrode sensor for detection of bisphenol A. *Ionics*, 25, 6, 2857–2864, 2019.
7. Bushra, R., Naushad, M., Adnan, R., Shahadat, M., Ansari, M., Ahmed, A., Electrical and optical properties of synthesized composite material polyaniline-Ti (IV) arsenophosphate. *Asian J. Chem.*, 27, 3, 1121–1124, 2015.
8. Shahadat, M., Kumar, S., Ismail, S., Ali, S.W., Ahammad S.Z., Clay-based adsorbents for the analysis of dye pollutants, *Applied Water Science Volume 1: Fundamentals and Applications*, pp. 163–197, John Wiley & Sons, Inc., 2021.
9. Khandegar, V. and Sahora, A.K., Electrocoagulation for the treatment of textile industry effluent–a review. *J. Environ. Manage.*, 128, 949–96.
10. Sahu, O. and Singh, N., Significance of bioadsorption process on textile industry wastewater, in: *The Impact and Prospects of Green Chemistry for Textile Technology*, pp. 367–416, Elsevier, Woodhead Publishing, 2018.
11. Good, K.D, VanBriesen, J.M., Power plant bromide discharges and downstream drinking water systems in Pennsylvania. *Environ. Sci. Technol.*, 51, 20, 11829–11838, 2017.
12. Kalyani, D., Telke, A., Phugare, S., Govindwar, S., Jadhav, J.P., Evaluation of the efficacy of a bacterial consortium for the removal of color, reduction of heavy metals, and toxicity from textile dye effluent. *Bioresour. Technol.*, 101, 165–173, 2010.
13. Central Pollution Control Board, Standards for effluents from textile industry, in: *The Environment Rules 1986*, pp. 1–16, 2010.
14. Ramamoorthy, S.K., Lavate, S.S., Periyasamy, A.P., Eco-friendly denim processing, in: *Handbook of Ecomaterials*, L.M.T. Martínez, O.V. Kharissova, B.I. Kharisov, (Eds.), pp. 1–21, Springer International Publishing, Cham, 2019.
15. Venkatesan, H. and Periyasamy, A.P., Eco-fibers in the textile industry, in: *Handbook of Ecomaterials*, L.M.T. Martínez, O.V. Kharissova, B.I. Kharisov, (Eds.), pp. 1–21, Springer International Publishing, Cham, 2017.

16. Periyasamy, A.P., Rwahwire, S., Zhao, Y., Environmental friendly textile processing, in: *Handbook of Ecomaterials*, L.M.T. Martínez, O.V. Kharissova, B.I. Kharisov, (Eds.), pp. 1–38, Springer International Publishing, Cham, 2018.
17. Khan, R., Bhawana, P., Fulekar, M.H., Microbial decolorization and degradation of synthetic dyes: A review. *Rev. Environ. Sci. Biotechnol.*, 12, 1, 75–97, 2012.
18. Mani, S. and Bharagava, R.N., Exposure to crystal violet, its toxic, genotoxic and carcinogenic effects on environmental and its degradation and detoxification for environmental safety. *Rev. Environ. Contam. Toxicol.*, 237, 71–104, 2016.
19. Chen, Y., Truong, V.N.T., Bu, X., Xie, G., A review of effects and applications of ultrasound in mineral flotation. *Ultrason. Sonochem.*, 60, 104739, 2020.
20. Bizuneh, A., Textile effluent treatment & decolorization techniques. *Chem. Bulg. J. Sci. Educ.*, 21, 434–456, 2012.
21. Miralles-Cuevas, S., Oller, I., Agüera, A., Combination of nanofiltration and ozonation for the remediation of real municipal wastewater effluents: Acute and chronic toxicity assessment. *J. Hazard. Mater.*, 323, 442–451, 2016.
22. Gosavi, V.D. and Sharma, S., A general review on various treatment methods for textile wastewater. *J. Environ. Sci. Comput. Sci. Eng. Technol.*, 3, 29–39, 2014.
23. Wang, L.K., Shammas, N.K., Williford, C. *et al.*, Evaporation processes, in: *Advanced Physicochemical Treatment Process*, L.K. Wang, N.K. Shammas, Y.-T. Hung, (Eds.), pp. 549–579, Humana Press, Totowa, 2006.
24. Pletcher, D. and Walsh, F.C., Water purification, effluent treatment and recycling of industrial process streams, in: *Industrial Electrochemistry*, D. Pletcher, and F.C. Walsh, (Eds.), pp. 331–384, Springer, Dordrecht, 1993.
25. Srithar, K. and Mani, A., Studies on solar flat plate collector evaporation systems for tannery effluent (soak liquor). *J. Zhejiang Univ. Sci. A*, 7, 1870–1877, 2006.
26. Sarayu, K. and Sandhya, S., Current technologies for biological treatment of textile wastewater-a review. *Appl. Biochem. Biotechnol.*, 167, 645–661, 2012.
27. Ranganathan, K., Karunagaran, K., Sharma, D.C., Recycling of wastewaters of textile dyeing industries using advanced treatment technology and cost analysis—Case studies. *Resour. Conserv. Recycl.*, 50, 306–318, 2007.
28. Mauskar, J.M., *Advance Methods for Treatment of Textile Industry Effleunts*, pp. 1–137, Central Pollution Control Board, India, 2007.
29. Liu, H., Zhao, X., Qu, J., Electrocoagulation in water treatment, in: *Electrochemistry Environmental*, C. Comninellis, and G. Chen (Eds.), pp. 245–262, Springer, New York, 2010.
30. Van der Bruggen, B., Canbolat, C.B., Lin, J., Luis, P., The potential of membrane technology for treatment of textile wastewater, in: *Sustainable Membrane Technology Water Wastewater Treatment*, A. Figoli, and A. Criscuoli, (Eds.), pp. 349–380, Springer, Singapore, 2017.

31. Sahu, O., Mazumdar, B., Chaudhari, P.K., Treatment of wastewater by electrocoagulation: A review. *Environ. Sci. Pollut. Res.*, 21, 2397–2413, 2014.
32. Yuksel, E., Eyvaz, M., Gurbulak, E., Electrochemical treatment of colour index reactive orange 84 and textile wastewater by using stainless steel and iron electrodes. *Environ. Prog. Sustain. Energy*, 32, 60–68, 2013.
33. Khandegar, V. and Saroha, A.K., Electrochemical treatment of textile effluent containing acid red 131 dye. *J. Hazard. Toxic Radioact. Waste*, 18, 38–44, 2014.
34. Patel, U.D., Ruparelia, J.P., Patel, M.U., Electrocoagulation treatment of simulated floor-wash containing reactive black 5 using iron sacrificial anode. *J. Hazard. Mater.*, 197, 128–136, 2011.
35. Bayramoglu, M., Kobya, M., Can, O.T., Decolorization of reactive dye solutions by electrocoagulation using aluminum electrodes. *Ind. Eng. Chem. Res.*, 42, 3391–3396, 2003.
36. Polgumhang, S., Tongdaung, W., Phalakornkule, C., Performance of an electrocoagulation process in treating direct dye: Batch and continuous upflow processes. *World Acad. Sci. Eng. Technol.*, 3, 267–272, 2009.
37. Sorkhabi, H.A., Kasiri, M., Daneshvar, N., De-colourisation of dye containing acid red 14 by electrocoagulation with comparartive investigation of different electrode connections. *J. Hazard. Mater.*, 112, 55–62, 2004.
38. Morão, A., Magrinho, M., Lopes, A., Gonçalves, I., Fernandes, A., Electrochemical degradation of C. I. Acid orange 7. *Dyes Pigments*, 61, 287–296, 2004.
39. Merzouk, B., Gourich, B., Madani, K. *et al.*, Removal of a disperse red dye from synthetic wastewater by chemical coagulation and continuous electrocoagulation. *A Comp. study. Desalination*, 272, 246–253, 2011.
40. Özyonar, F., Gökkuş, Ö., Sabuni, M., Removal of disperse and reactive dyes from aqueous solutions using ultrasound-assisted electrocoagulation. *Chemosphere*, 258, 127–235, Nov. 2020.
41. Shokouhi, S.B., Dehghanzadeh, R., Aslani, H., Shahmahdi, N., Activated carbon catalyzed ozonation (ACCO) of reactive blue 194 azo dye in aqueous saline solution: Experimental parameters, kinetic and analysis of activated carbon properties. *J. Water Process Eng.*, 35, 101188, 2020.
42. Rana, S., Gupta, N., Rana, R.S., Removal of organic pollutant with the use of rotating biological contactor. *Mater. Today Proc.*, 5, 4218–4224, 2018.
43. Khosravi, A., Karimi, M., Ebrahimi, H., Fallah, N., Sequencing batch reactor/nanofiltration hybrid method for water recovery from textile wastewater contained phthalocyanine dye and anionic surfactant. *J. Environ. Chem. Eng.*, 8, 103701, 2020.
44. Madhu, A. and Chakraborty, J.N., Developments in application of enzymes for textile processing. *J. Clean. Prod.*, 145, 114–133, 2017.
45. Ji, D. *et al.*, Green preparation of polyvinylidene fluoride loose nanofiltration hollow fiber membranes with multilayer structure for treating textile wastewater. *Sci. Total Environ.*, 754, 141–848, Feb. 2021.

46. Long, Q., Zhang, Z., Qi, G., Wang, Z., Chen, Y., Liu, Z.Q., Fabrication of chitosan nanofiltration membranes by the film casting strategy for effective removal of dyes/salts in textile wastewater. *ACS Sustain. Chem. Eng.*, 8, 6, 2512–2522, 2020. https://doi.org/10.1021/acssuschemeng.9b07026
47. Karisma, D., Febrianto, G., Mangindaan, D., Removal of dyes from textile wastewater by using nanofiltration polyetherimide membrane. *IOP Conf. Ser.: Earth Environ. Sci.*, 109, 1–6, 2017.
48. Zhang, L., Xu, L., Yu, H., Yao, P., Zhang, M., Guo, F., Yu, L., Capsaicin mimic-polyethyleneimine crosslinked antifouling loose nanofiltration membrane for effective dye/salt wastewater treatment. *J. Membr. Sci.*, 641, 119–923, 2022.
49. Rahimi, Z., Zinatizadeh, A.A., Zinadini, S., van Loosdrecht, M., A hydrophilic and antifouling nanofiltration membrane modified by citric acid functionalized tannic acid (CA-f-TA) nanocomposite for dye removal from biologically treated baker's yeast wastewater. *J. Environ. Chem. Eng.*, 9, 1, 104963, 2021.
50. Salahshoor, Z., Shahbazi, A., Maddah, S., Magnetic fielde influenced nanofiltration membrane blended by CSeEDTAemGO as multiefunctionality green modifier to enhance nanofiltration performance, efficient removal of Na2SO4/Pb2þ/RR195 and cyclic wastewater treatment. *Chemosphere*, 278, 130–379, 2021.
51. Halakarni, M., Mahto, A., Aruchamy, K., Mondal, D., Kotrappanavar, N., Developing helical carbon functionalized chitosan-based loose nanofiltration membranes for selective separation and wastewater treatment. *Chem. Eng. J.*, 417, 127911, 2021.
52. Dąbrowski, A., Adsorption—From theory to practice. *Adv. Colloid Interface Sci.*, 93, 1, 135–224, 2001.
53. Shireesha, M., Rao, P.K., Ali, M.D.H., Chaitanya, K.V., A review on effluent treatment of textile by biological and chemical methods. *Int. J. Eng. Technol. Sci. Res.*, 4, 2394–3386, 2017.
54. Hasan, M., Shenashen, M.A., Hasan, N., Znad, H., Salman, M.S., Awual, M.R., Natural biodegradable polymeric bioadsorbents for efficient cationic dye encapsulation from wastewater. *J. Mol. Liq.*, 323, 114587, 2021.
55. Kaur, M., Tewatia, P., Rattan, G., Singhal, S., Kaushik, A., Diamidoximated cellulosic bioadsorbents from hemp stalks for elimination of uranium (VI) and textile waste in aqueous systems. *J. Hazard. Mater.*, 417, 126–060, 2021. https://doi.org/10.1016/j.jhazmat.2021.126060
56. Khaleque, A. and Roy, D.K., Removing reactive dyes from textile effluent using banana fibre. *Int. J. Basic Appl. Sci.*, 16, 14–20, 2016.
57. Temesgen, F., Gabbiye, N., Sahu, O., Biosorption of reactive red dye (RRD) on activated surface of banana and orange peels: Economical alternative for textile effluent. *Surf. Interfaces*, 12, 151–159, 2018.

58. Jain, S.N., Tamboli, S.R., Sutar, D.S., Jadhav, S.R., Marathe, J.V., Shaikh, A.A., Prajapati, A.A., Batch and continuous studies for adsorption of anionic dye onto waste tea residue: Kinetic, equilibrium, breakthrough and reusability studies. *J. Clean. Prod.*, 252, 119778, 2020.
59. Narasaiah, B.P. and Mandal, B.K., Remediation of azo-dyes based toxicity by agro-waste cotton boll peels mediated palladium nanoparticles. *J. Saudi Chem. Soc.*, 24, 267–281, 2020.
60. Vasu, D., Kumar, S., Walia, Y.K., Removal of dyes using wheat husk waste as a low-cost adsorbent. *Environ. Claims J.*, 32, 67–76, 2020.
61. Mathew, M.L., Gopalakrishnan, A., Aravindakumar, C.T., Aravind, U.K., Low-cost multi-layered green fiber for the treatment of textile industry wastewater. *J. Hazard. Mater.*, 365, 297–305, 2019.
62. Jawad, A.H., Abdulhameed, A.S., Mastuli, M.S., Acid-factionalized biomass material for methylene blue dye removal: A comprehensive adsorption and mechanism study. *J. Taibah Univ. Sci.*, 14, 305–313, 2020.
63. Report on assessment of pollution from textile dyeing units in Tirupur, Tamil Nadu and measures taken to achieve zero liquid discharge. 2014-15 Central Pollution Control Board Zonal Office (South), Bengaluru.

7

Photocatalytic Activity of Green Mixed Matrix Membranes for Degradation of Anionic Dye

Oladipo, Gabriel Opeoluwa[1], Alayande, Samson Oluwagbemiga[2*], Ogunyinka Opeyemi O.[1], Akinsiku, Anuoluwa Abimbola[3], Akinsipo-Oyelaja, Oyesolape Basirat[4], Ofudje Edwin Andrew[5], Bolarinwa Hakeem S.[6], Akinlabi, Akinola Kehinde[7] and Msagati, Titus. A.M.[8]

[1]*Department of Science Laboratory Technology, D.S Adegbenro ICT Polytechnic, Itori-Ewekoro, Ogun State, Nigeria*

[2]*Department of Industrial Chemistry, First Technical University, Ibadan, Oyo State, Nigeria*

[3]*Department of Chemistry, Covenant University, Sango Ota, Ogun State, Nigeria*

[4]*Department of Chemical Sciences, Tai Solarin University of Education, Ijagun, Ogun State, Nigeria*

[5]*Department of Chemical Sciences, Mountain-Top University, Mowe, Nigeria*

[6]*Department of Physics, Electronics and Earth Sciences, Fountain University, Osogbo, Nigeria*

[7]*Department of Chemistry, Federal University of Agriculture, Abeokuta, Ogun State, Nigeria*

[8]*Nanotechnology and Water Sustainability Research Unit, College of Science Engineering and Technology, University of South Africa, Florida, South Africa*

Abstract

Anionic dye is a notable constituent of textile effluent, which renders water unsafe for human and animal use. A notable approach to mitigating effluent is the use of membranes. In this study, a mixed matrix technique was adopted for the preparation of composite membranes. The composite membranes consist of crumb rubber filled with nanoparticles, respectively. The membranes were characterized by Fourier transform spectroscopy, scanning electron microscopy coupled with energy dispersive x-ray, thermogravimetric analyser, and drop shape analyzer.

**Corresponding author*: samson.alayande@tech-u.edu.ng

Shahid-ul-Islam, Abid Hussain Shalla and Mohammad Shahadat (eds.) Green Chemistry for Sustainable Water Purification, (157–178) © 2023 Scrivener Publishing LLC

The photocatalytic activities of the mixed matrix membranes were investigated with anionic dye in the UV/visible region. Photocatalytic activity of composite membranes showed high degradation with an apparent rate constant. This study presents a rubber-based membrane for wastewater treatment.

***Keywords*:** Mixed matrix membranes, nanoparticles, natural crumb rubber, titanium (IV) oxide, anionic dye

7.1 Introduction

Wastewater treatment is a major channel to achieve safe water and eliminate surface water contamination. Membrane technology has recorded huge success in the field of wastewater treatment and water purification [1]. The first membrane was reported in 1960 by Loeb and Sourirajan, which was a cellulose acetate membrane. Twenty years later, a thin film composite reverse osmosis membrane was developed [1]. These developments have provoked global attention from researchers and industrialists to improve membrane performance and properties. Since then numerous inorganic, organic (synthetic and natural polymers) membranes have been reported for several applications [1]. In recent years, the interest in natural polymer has escalated due to their biodegradability and non-toxicity. It is worthy to note that properties, namely cost-effectiveness, hydrophilicity, semipermeability, porosity, film-forming ability, and good transport characteristics have provoked the use of natural polymer for membrane applications. Natural polymers such as chitin, chitosan, and cellulose are commonly used as membranes. Several efforts are being made to improve the performance of natural-based membranes, worthy efforts include incorporating additives such as organic and/or inorganic into the natural polymer [2]. This modification of natural polymers is critical for developing high-performance membranes. The incorporation of additives into natural polymers is classified as a mixed matrix membrane [3]. For example, cellulose acetate membrane was doped with TiO_2 nanoparticles, thermal stability and water permeation of the mixed matrix membrane were enhanced [4]. The incorporation of nanochitin whiskers in polyvinylidene fluoride membrane was found to improve the mechanical strength, as well as water permeability and antifouling properties [2]. The membrane of this nature is tailor-made for specific pollutants in wastewater treatment.

Several tailor-made mixed matrix membranes (MMM) have been reported in water/wastewater treatment. For instance, Venkatanarasinihan and Raghavachari, [5] reported the use of epoxidized natural rubber/Fe_3O_4

membrane for removal of oil. Membrane-based on TiO_2/ZnO-cellulose composite was reported effective for photocatalytic degradation of phenol [6], methylene blue, humic acid [7], rhodamine B [8], and methyl orange [9]. Zhao *et al.* [10] observed that the addition of TiO_2 enhanced membrane functionality through adsorption, which was simultaneously accompanied by degradation of pollutant dye. The advantages of MMM for effluent treatment are high permeance performance for removal of pollutants particularly organic compounds, membrane recovery, and composite formation.

A notable constituent of the tailor-made membrane is a photocatalyst, which is incorporated photodegradation of pollutants, namely organic and inorganic, and disinfestation of wastewater. The most popular photocatalyst is TiO_2. A viable route for the enhancement of the photocatalytic property of TiO_2 under visible light is the use of doping technology. The dopants, such as metallic ions or nonmetallic ions, were introduced into TiO_2 lattice during synthesis [11]. Titanium dioxide is the most widely used product of titanium, about 95% of titanium ore is currently processed into TiO_2. The global market value is estimated at around 17 billion US dollars, an annual increment of eight million tonnes. Its application cut across various sectors; wastewater treatment inclusive. In wastewater treatment, TiO_2 is used for disinfestation, microbial inhibition, and photocatalytic degradation. Properties, such as high photocatalytic activity in the UV region, affordability, nontoxic nature, chemical stability, availability, and strong oxidizing power favor its use in wastewater treatment processes. In order to shift photocatalytic activity from UV to visible region, which requires lower energy, doping with impurities usually metallic/nonmetallic substances have yielded huge success with enhanced efficiency, particularly for metallic dopants [11].

On the other hand, natural rubber is also known as cis-1,4-poly(isoprene), natively obtained from the tree of *Hevea brasiliensis.* It is nontoxic, biomaterial, flexible, biodegradable, affordable, readily available, and sourced from renewable feedstock. When concentrated or coagulated with acetic acid and dried forms crumb rubber [12]. Further subjected to processing into rubber sheets, vulcanization, and compounding. Conventional, fillers are added to rubber to improve mechanical, electrical, thermal, and chemical properties. The process leads to the rubber reinforcement effect. The effect has been reported for enhanced stiffness, cross-linking of rubber and filler, in case of multicomponent filler system; filler-filler pairing. Two notable hysteretic mechanisms are Payne and Mullins's effects [13]. The latest concept involves the use of nanomaterials as fillers. This is largely due to the large interfacial area provided with the rubber matrix, therefore, giving rise to high interaction between rubber and filler network [13].

In the present study, natural crumb rubber, a natural polymer was utilized as a support for TiO_2, Ag-TiO_2, and Ag/Zn-TiO_2 nanoparticles for the preparation of Mixed Matrix Membrane for photocatalytic degradation of methyl orange dye under visible light.

7.2 Materials and Methods

7.2.1 Materials

Natural rubber latex was obtained from the Rubber Research Institute of Nigeria (RRIN), Benin City, Nigeria. All chemicals (2-propanol, titanium tetraisopropoxide, zinc trioxonitrate (V) hexahydrate concentrated nitric acid, and silver trioxonitrate (V)) used were purchased from Sigma Aldrich, South Africa. Q-Millipore purification system was used deionizing water used.

7.2.2 Methods

7.2.2.1 Synthesis of TiO_2 Nanoparticles

TiO_2, Ag-doped TiO_2 and Ag-Zn co-doped TiO_2 nanopowders were prepared using our previous method [11, 14].

7.2.2.2 Preparation of Natural Rubber Composites

20 g of dry crumb rubber was mixed with a required amount of NR (without nanoparticles), TiO_2, Ag doped TiO_2 and Ag-Zn co-doped TiO_2 nanoparticle. The samples were calculated by wt% based on the amount of the dopant added to crumb rubber using Rheomixer. The RheoDrive 7 was connected to Polysoft OS software where operation conditions were set (mixing temperature, 200°C and time, 20 min). The membranes were extruded out in a flat shape with 40 mm thickness. The following samples were prepared NR, NR-4% TiO_2, NR-4% Ag-TiO_2, NR-4%Ag-Zn co-doped TiO_2, NR-6% TiO_2, NR-6% Ag-TiO_2, and NR-6% Ag-Zn co-doped TiO_2 nanoparticles.

7.2.3 Analysis

7.2.3.1 Micrograph Analysis

The phase morphology, particle size distribution, and elemental composition of nanoTiO_2 and doped nanoTiO_2 were examined using SEM/EDS (model JSM-IT300 JEOL Ltd., Japan).

A small amount or portion of dried samples was placed on the top of the SEM-EDX stub using carbon tape. The samples were coated with a gold thin layer of 10 μm thickness via a Hummer X sputtering coater of Quorum (model Q150R ES) [15].

7.2.3.2 Structural Analysis

The FT-IR spectra of composite membranes were analyzed at room temperature using a PerkinElmer FT-IR spectrometer, Frontier. A portion of the membrane was mounted on the ATR sample holder, in a spectra range of 4000 to 400 cm [16].

7.2.3.3 Thermal Analysis

TGA of the samples was carried out with TGA thermal analyzer (TGA 5500 Discovery series). The heating rate was 10°C/min with a flow of nitrogen gas at 50 mL/min heated from 50°C to 900°C [17].

7.2.3.4 Wetting Analysis

The composite membrane was mounted on the sample holder using double-sided foam tape. A test fluid was chosen and placed in the syringe. Water was used as a test fluid even though other fluids may be required for surface energy analysis. A video image of the sample was obtained by adjusting focus and lighting. The needle position was adjusted so that the tip is visible in the image [18].

7.2.3.5 Photocatalytic Performance

The photocatalytic activity of the membranes on methyl orange dye was investigated under a solar simulator (HAL-320 F.S, ASAHI SPECTRA). An initial concentration of 4 ppm of methyl orange dye was used for the degradation experiments. A portion of composite membrane mixed with 50 mL of methyl orange solution was stirred at 100 rpm for 30 min without solar irradiation. This step is essential to ascertain adsorption-desorption equilibrium. At 30-min intervals, a Millipore Millex-LCR hydrophilic filter was used to remove the membrane while 3 mL of the suspensions were pipetted. These solutions were subjected to further analysis mainly absorption measurement. PerkinElmer Lamda 650 S UV–Vis spectrophotometer was used in this study. To irradiate the solution with visible light, a Compact Xenon lamp was positioned at 80 mm away from the dye solution. The concentration of degraded methyl orange dye solution was determined from the calibration one plotted with six standard solutions. The degradation efficiency was estimated as follows:

$$\text{Degradation efficiency}\ (\%) = \frac{Co - Ct}{Co} \times 100$$

The degradation of dye was estimated from the ratio (C/C_o) concerning irradiation time. The plot of $\ln(C_o/C)$ against irradiation time was linear, then the slope is rate constant (k) can be extrapolated for each sample used for the photocatalytic degradation [19].

$$\ln\left({Co}/{C} \right) = k_{app} t$$

where C_o is the initial concentration of methyl orange dye, k_{app} is apparent rate constant while C is the concentration after irradiation at a time interval, t.

7.3 Results and Discussion

7.3.1 Fourier Transform Infrared Spectroscopy of Composites Membranes

Chemical molecular vibrations of the natural rubber and composite membranes were confirmed by the FT-IR technique and shown in Figures 7.1a-b. The composites show similar spectra, which include C-H bands notable for methylene symmetric and asymmetric stretching at 2852 cm and 2918 cm, respectively. Another peak was connected with asymmetric stretching vibration of methyl, a group notable at 2958 cm. Also, the vibration of a hydrogen-bonded C=O group band was notable at 1665 cm [20, 21]. The C=C stretching vibration at 1675 cm was observed in the natural rubber matrix. The presence of inorganic additives may be confirmed in the fingerprint region, which is noted for several interferences.

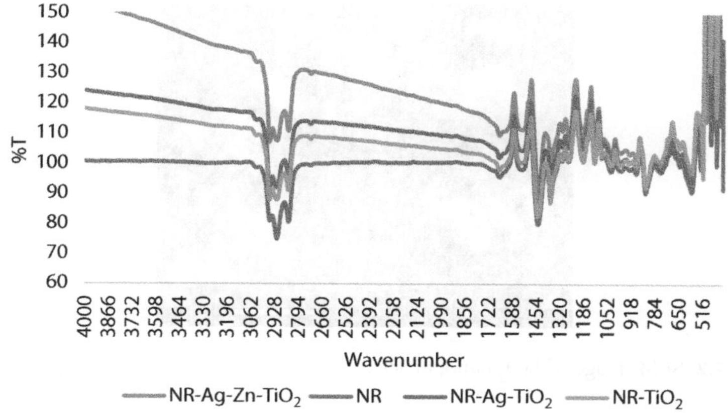

(a)

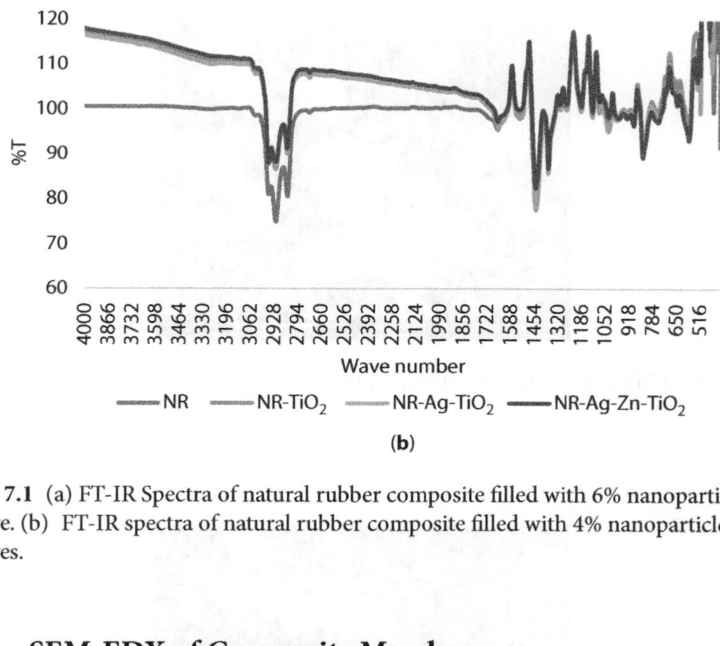

(b)

Figure 7.1 (a) FT-IR Spectra of natural rubber composite filled with 6% nanoparticles additive. (b) FT-IR spectra of natural rubber composite filled with 4% nanoparticles additives.

7.3.2 SEM-EDX of Composite Membranes

The observed SEM micrographs of TiO_2, doped or co-doped TiO_2 are presented in Figures 7.2, 7.3, and 7.4. The particles are tetragonal in shape, and the average sizes are found in the range of 50 to 170 nm at different magnification. The variation of sizes and morphology is due to particle agglomeration. It can also be observed that the Ag dopant is seen on the surfaces of the particle as a dotted black spot.

Figure 7.2 SEM image of TiO_2 nanoparticle.

Figure 7.3 SEM image of Ag-TiO_2 nanoparticle.

Figure 7.4 SEM image of Ag-Zn-TiO_2 nanoparticle.

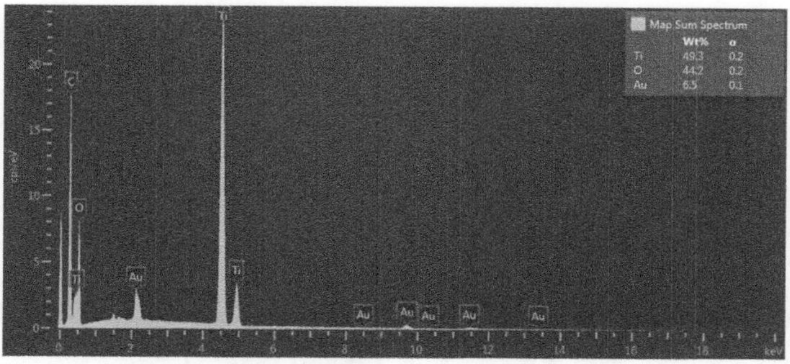

Figure 7.5 EDS Mass spectrum of TiO_2 nanoparticle.

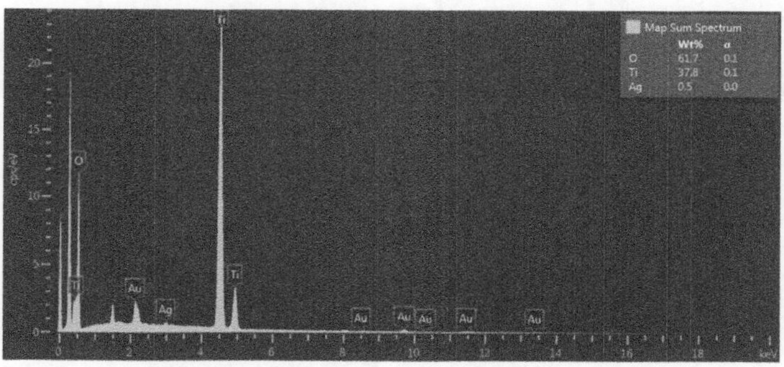

Figure 7.6 EDS mass spectrum of Ag-TiO_2 nanoparticle.

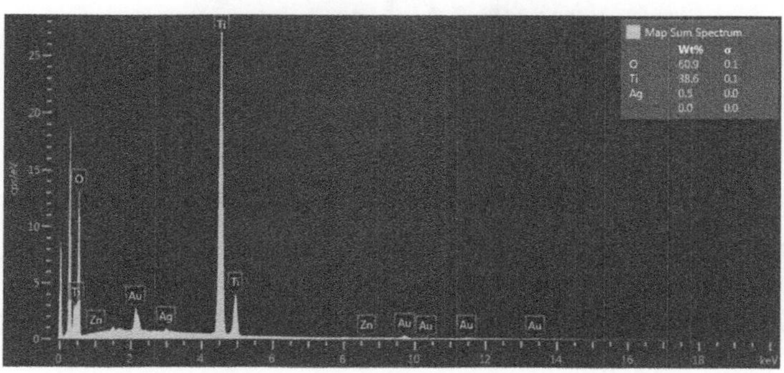

Figure 7.7 EDX mass spectrum of Ag-Zn-TiO_2 nanoparticle.

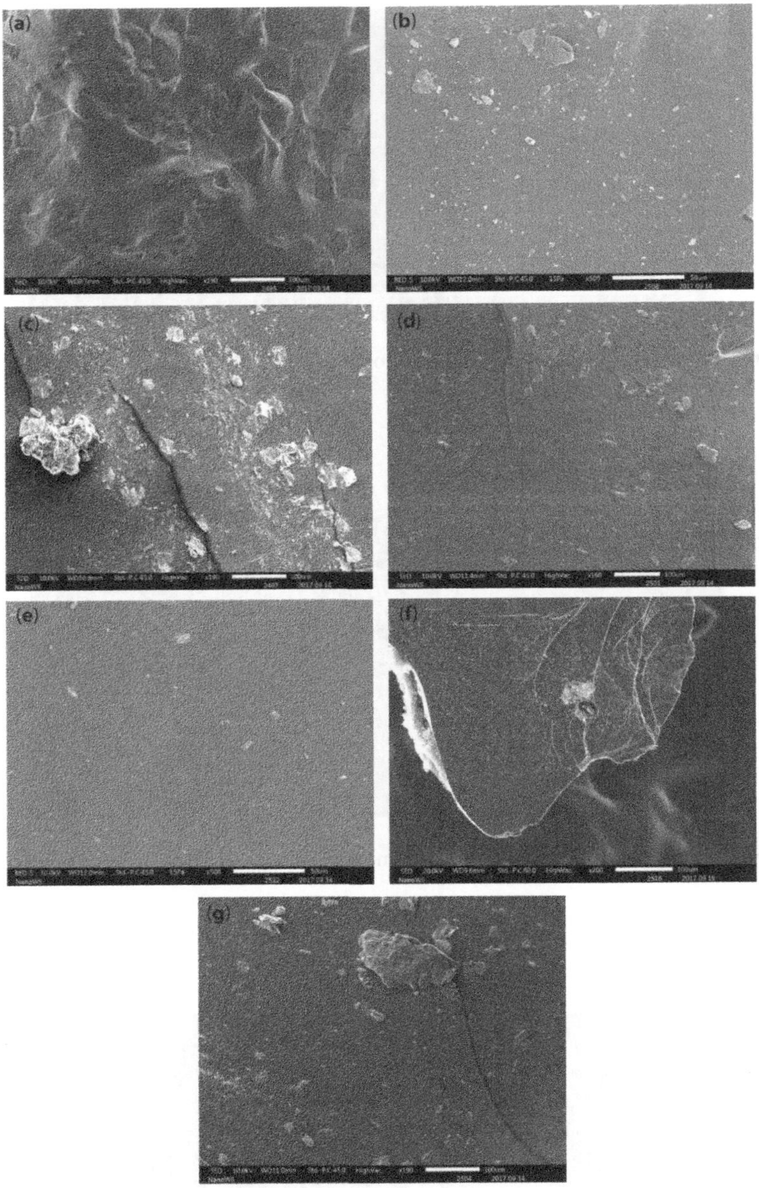

Figure 7.8 (a) SEM image of natural rubber (NR), (b) SEM image of NR-4% TiO_2, (c) SEM image of NR-6% TiO_2, (d) SEM image of NR-4% Ag-TiO_2, (e) SEM image of NR-6% Ag-TiO_2, (f) SEM image of NR-4% Ag-Zn-TiO_2, (g) SEM image of NR-6% Ag-Zn-TiO_2.

The examination of nanopowder by EDX technique confirms the presence of Ti, O, Ag, and Zn as well as impurities from the gold coating and carbon tape as shown in Figures 7.5, 7.6, and 7.7. From the weight proportions of the elements identified, one can observe a high oxygen content of 61.7% and 60.9% for doped and co-doped TiO_2 nanoparticles, but somewhat lower (44.2%) for undoped TiO_2 nanoparticles. The high oxygen content for doped and codoped TiO_2 may be from the salts used during the doping synthesis. However, Zn wt% content might be below the sensitivity of the instrument.

Figure 7.8a-g show the SEM micrographs of NR, 4% Ag doped, 6% Ag/Zn co-doped TiO_2 rubber composite membranes. The nanoparticles were found to be dispersed on the surface of the natural rubber (NR). However, there was some large clusters of the particles on the surface of NR, which indicate the uneven distribution of the particles due to particles' agglomeration. This might hinder the availability of the nanoparticles for photogeneration of more electron holes.

EDX of the micrograms shows carbon, titanium, oxygen, silver, and zinc. In the elements identified, carbon predominates, while oxygen, silver, and zinc are distributed evenly.

7.3.3 Thermogravimetric Analysis of Composite Membranes

The thermogram of the rubber composite membranes is presented in Figure 7.9. Thermal degradation of composite membranes begins at 300°C. All composite membranes showed a single-stage thermal degradation. The thermal decomposition was notable slightly below 400°C in all the composites membranes; however, there was little variation in the two composites, which was not proportional to the nanoparticles content. One of the challenges with rubber compounding is even distribution of compounding ingredients on the rubber surfaces, this has limited the use of natural rubber. The addition of nanoparticles to NR resulted in to increase in thermal stability as a result of the nanoparticles [11]. The highest thermal stability was exhibited by NR-4%Ag-Zn-TiO_2 composite membrane with about 70.25% weight loss while NR showed the lowest thermal stability with 99.32% weight loss.

7.3.4 Contact Angle Measurement of Composite Membranes

Table 7.1 shows the water contact angle of all the rubber composite membranes. The values ranged from 76.3 to 85.8°. The NR is observed to have the lowest value in comparison with others. Water contact angle below

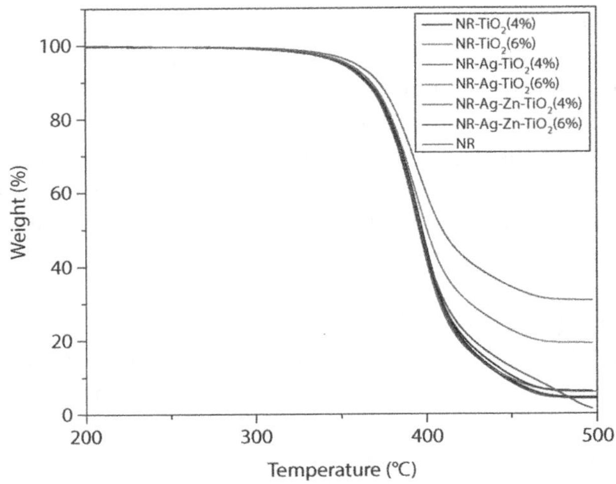

Figure 7.9 Thermograms of NR and composite membranes.

Table 7.1 Contact angle measurement.

Composite	**Contact angle**
NR	76.3 ± 5.4
NR-TiO_2 (4%)	85.8 ± 16.0
NR-TiO_2 (6%)	84.2 ± 13.0
NR-Ag-TiO_2 (4%)	82.7 ± 6.0
NR-Ag-TiO_2 (6%)	81.8 ± 6.3
NR-Ag-Zn-TiO_2 (4%)	81.2 ± 9.1
NR-Ag-Zn-TiO_2 (6%)	80.4 ± 5.4

90° describes favorable wetting of the surface, such fluid such as water will spread over a large area on the membrane. This refers to the hydrophilic property of the membrane. When water contact angles > 90°, this denotes hydrophobic property [22]. This means water will minimize its contact with the membrane surface. The wetting of such membrane will be unfavourable [22]. The membranes are hydrophilic which tend toward hydrophobicity. Such membranes are applicable for wastewater treatment; these materials can be applied without membrane dissolution in pollutants.

7.3.5 Photodegradation of Composite Membranes

The adsorption of methyl orange dye on NR filled with 4/6% TiO_2 doped/undoped nanoparticles investigated under solar irradiation at pH 4.1 is shown in Figure 7.10 and illustrated by absorption spectra (Figures 7.11–7.17). It was observed that the 6% NR-Ag/Zn-TiO_2 membrane showed the highest adsorption and/or degradation of methyl orange at 33%. The efficiency of 26% and 17% was obtained for TiO_2 membranes doped and undoped with Ag respectively for the same dye. However, 14%, 25%, and 21% degradation are obtained for NR with 4% additive membranes respectively. The absorption peak at 274 nm signifies π-π^* transition notable in the benzene ring. This peak increases as the irradiation time increases. These might be attributed to the adsorption capacity of membranes coupled with slow photodegradation of methyl orange dye by catalysts. This might suggest that the azo group (-N=N) of methyl orange dye is decomposed to another benzene derivative rather than complete mineralization. The lowest adsorption (8%) is observed for natural rubber. The low concentration of catalysts in the entire portion of the composite membrane may be responsible for the lower rate of photodegradation compared to the previous success achieved with the nanoparticles [11].

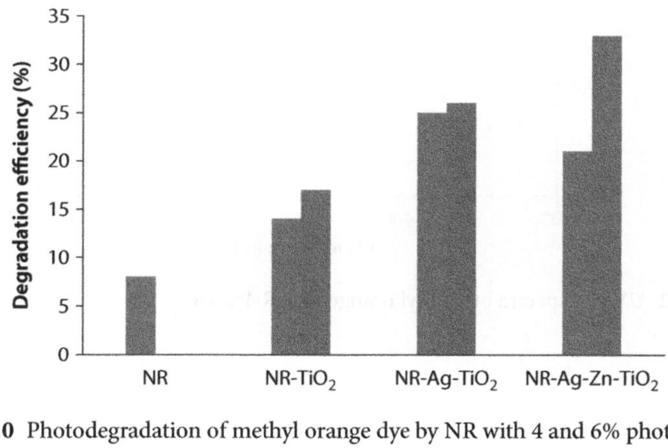

Figure 7.10 Photodegradation of methyl orange dye by NR with 4 and 6% photocatalyst at pH 4.1.

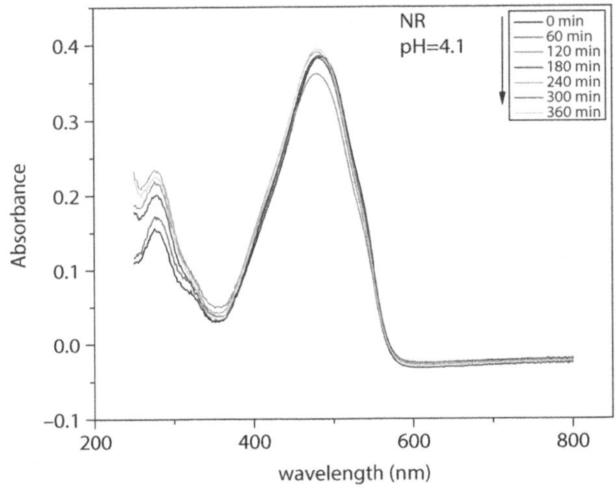

Figure 7.11 UV-Vis spectra of methyl orange by NR.

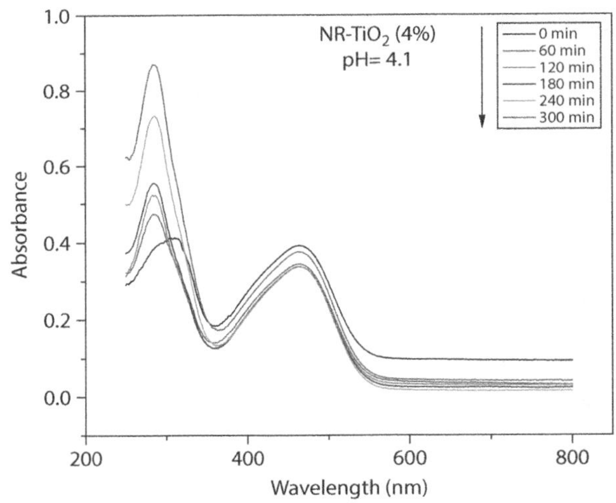

Figure 7.12 UV-Vis spectra of methyl orange by NR 4% TiO_2.

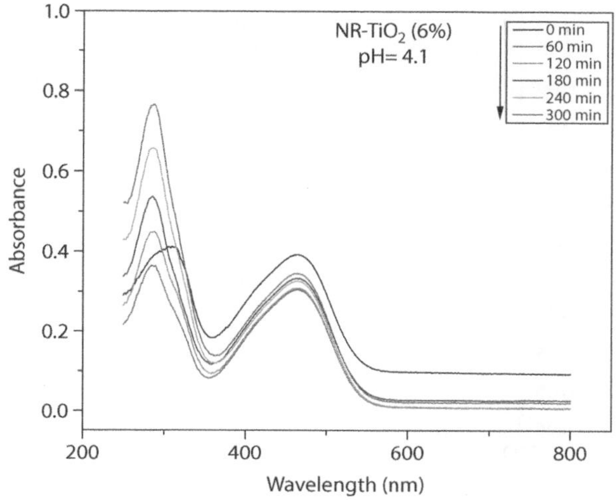

Figure 7.13 UV-Vis spectra of methyl orange by NR 6% TiO_2.

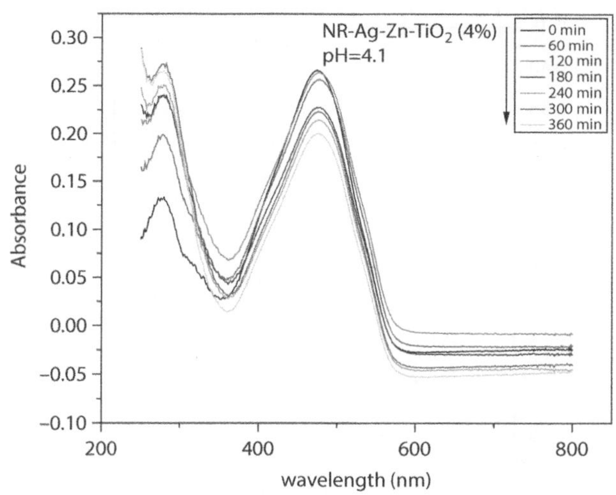

Figure 7.14 UV-Vis spectra of methyl orange by NR 4% Ag-TiO_2.

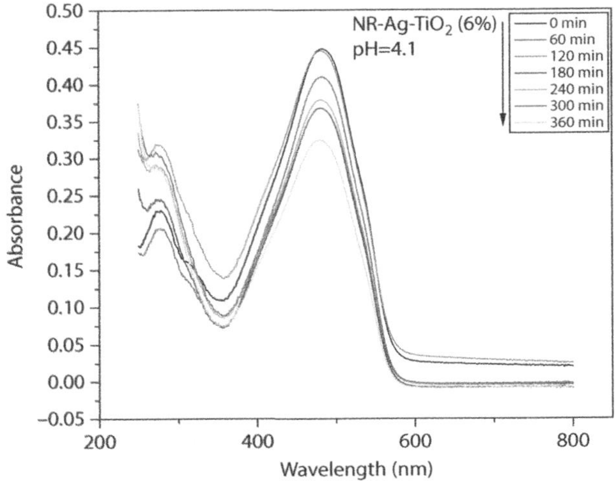

Figure 7.15 UV-Vis spectra of methyl orange by NR 6% Ag-TiO_2.

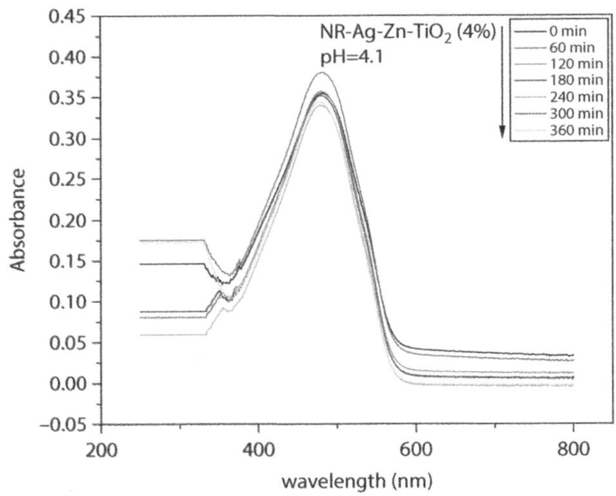

Figure 7.16 UV-Vis spectra of methyl orange by NR 4% Ag-Zn-TiO_2.

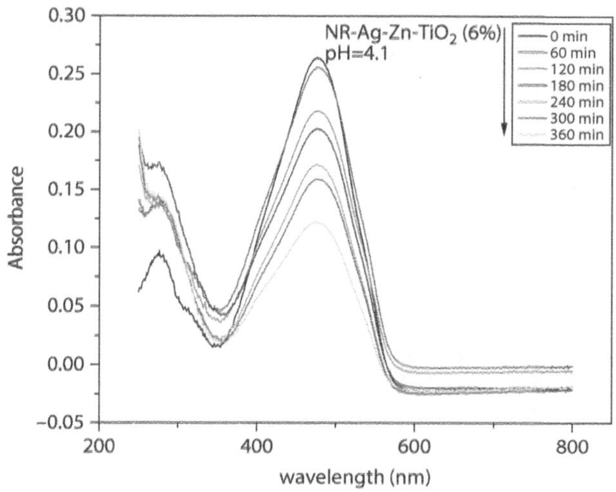

Figure 7.17 UV-Vis spectra of methyl orange by NR 6% Ag-Zn-TiO_2.

The degradation of methyl orange using rubber membranes was calculated using Langmuir-Hinshelwood adsorption kinetic model. In Figure 7.18, a sharp concentration decrease is observed for all the composites membranes at the initial stage of photocatalysis. This becomes relatively slow and steady as the irradiation time increases. However, Ag and/or Zn codoped TiO_2 membranes showed different trends, and these accounted for their enhanced percentage photodegradation. The different deficiencies may be attributed to the location of the TiO_2 nanoparticle on the rubber matrix and their contact with the dye and visible light. The concentration of dye diminished during reaction linearly with the contact time, indicating degradation. The percentage degradation at 360 min was 52% in this study. There was reduction in efficiency of the photocatalysts and increased contact time compared to our previous study using doped and undoped TiO_2 nanoparticles [11]. The reason for these is mainly due to impregnation in the host, reducing surface area, and activity of photocatalysts.

In Figure 7.19, the L-H model used to study the degradation rates of the dye showed kinetics properties. Kinetics of heterogeneous catalytic processes has been widely studied by pseudo-first-order kinetics and Langmuir–Hinshelwood (L–H) models [19]. The apparent first-order rate constant for the composite membranes filled with nanoparticles was also monitored at pH 4.1 and shown in Figure 7.19. NR with 4/6% TiO_2 nanoparticle was showed poor photodegradation and their apparent first-order rate constants were approximately zero. The highest photodegradation

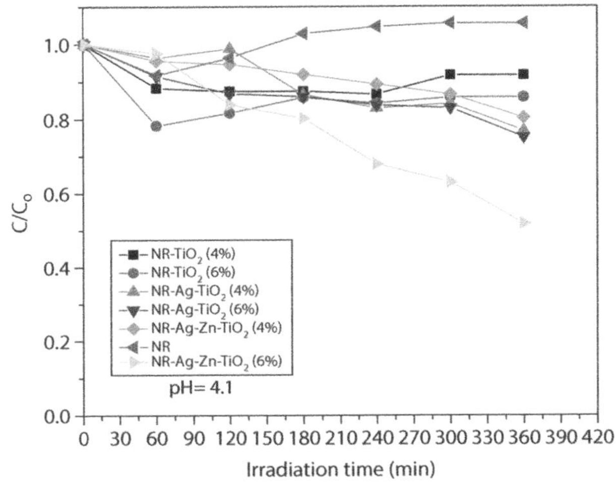

Figure 7.18 Photocatalytic degradation of methyl orange dye by NR and composites membranes.

rate was observed at 0.00182 min for NR with 6% Ag-Zn-TiO_2 nanoparticle followed by NR with 4% Ag-TiO_2 (0.00076 min). Meanwhile, the apparent-first-order rate constant of NR-4%Ag/Zn-TiO_2 and NR-6% Ag-TiO_2 membranes are 0.00054 and 0.00065 min, respectively. This simply revealed the

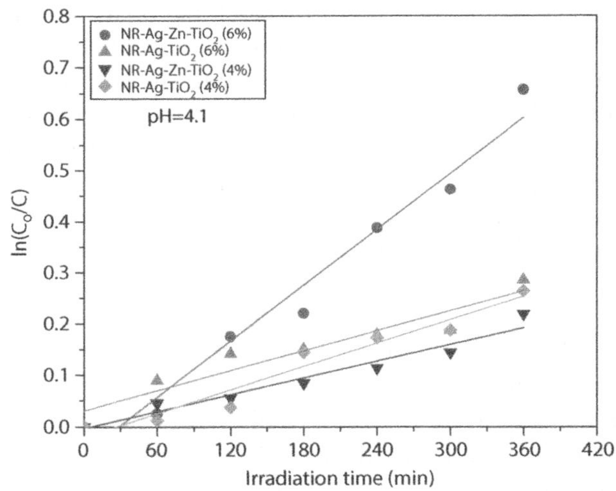

Figure 7.19 The degradation rates of methyl orange dye by NR composite membranes.

rate at which the dye is degraded and consequently, the highest rate constant gives the highest percentage degradation.

7.4 Conclusion

The result shows the effects of metal dopants on TiO_2 nanoparticles and subsequent immobilization on natural rubber for photodegradation of simulated textile effluent. The thermal stability of the NR membranes was improved with impregnation with photocatalysts. The mixed matrix membranes showed relatively reduced photodegradation of the organic. However, the highest degradation is attained with a composite membrane filled with 6% Ag-Zn-TiO_2 nanoparticles. The pseudo-first-order and apparent-first-order rate constant ranging from 0.00054 to 0.00182 min are appropriate models for the photodegradation process.

References

1. Asmaa, E., Nady, N., Bassyouni, M., El-Shazly, A., Metal organic framework based polymer mixed matrix membranes: Review on applications in water purification membranes. *Membranes*, 9, 88, 1–31, 2019.
2. Qin, A., Li., X., Zhao, X., Liu, D., He, C., Preparation and characterization of nano-chitin whisker reinforced PVDF membrane with excellent antifouling property. *J. Membr. Sci.*, 480, 1–10, 2015.
3. Alamaria, A.H.M., Nawawi, M.G.M., Zamrud, Z., Sago/PVA blend membranes for the recovery of ethyl acetate from water. *Arab. J. Chem.*, 20, 1517–1522, 2015.
4. Abedini, R., Mousavi, S.M., Aminzadeh, R., A novel cellulose acetate (CA) membrane using TiO_2 nanoparticles: Preparation, characterization, and permeation study. *Desalination*, 277, 40–45, 2011.
5. Venkatanarasimhan, S. and Raghavachari, D., Epoxidized natural rubber–magnetite nanocomposites for oil spill recovery. *J. Mater. Chem. A*, 1, 868–876, 2013.
6. Zeng, J., Liu, S., Zhang, L., TiO_2 immobilized in cellulose matrix for photocatalytic degradation of phenol under weak UV light irradiation. *J. Phys. Chem. C*, 114, 7806–7811, 2010.
7. Bai, H., Liu, Z., Sun, D.D., Hierarchically multifunctional TiO_2 nano-thorn membrane for water purification. *Chem. Commun.*, 46, 6542–6544, 2010.
8. Li, C., Liu, Q., Shu, S., Xie, Y., Zhao, Y., Chen, B., Preparation and characterization of regenerated cellulose/TiO2/ZnO nanocomposites and its photocatalytic activity. *Mater. Lett.*, 117, 234–236, 2014.

9. Zhang, J., Liu, W., Wang, P., Qian, K., Photocatalytic behavior of cellulose-based paper with TiO2 loaded on carbon fibers. *J. Environ. Chem. Eng.*, 1, 175–182, 2013.
10. Zhao, K., Feng, L., Lin, H., Fu, Y., Lin, B., Cui, W., Li, S., Wei, J., Adsorption and photocatalytic degradation of methyl orange imprinted composite membranes using TiO_2/calcium alginate hydrogel as matrix. *Catal. Today*, 236, 127–134, 2014.
11. Oladipo, G.O., Akinlabi, A.K., Alayande, S.O., Msagati, T.A.M., Nyoni, H.H., Ogunyinka, O.O., Synthesis, characterization and photocatalytic activity of silver and zinc codoped TiO_2 nanoparticle for photodegradation of methyl orange dye in aqueous solution. *Can. J. Chem.*, 97, 9, 642–650, 2019.
12. Akinlabi, A.K., Oladipo, G.O., Alayande, S.O., Msagati, T.A.M., Nyoni, H., Ogunyinka, O.O., Mosaku, A.M., Synthesis of titanium (IV) oxide rubber composite membrane for photodegradation of methyl orange dye. *J. Chem. Soc Nigeria*, 43, 1, 217 – 230, 2018.
13. Liliane, B., Natural rubber nanocomposites: A review. *Nanomaterials*, 9, 12, 1–21, 2019.
14. Perumal, S., Sambandam, C.G., Prabu, K.M., Ananthakumar, S., Synthesis and characterization studies of nano TiO_2 prepared via sol-gel method. *Int. J. Res. Eng. Technol.*, 3, 4, 651–657, 2014.
15. Aware, D.V. and Jadhav, S.S., Synthesis, characterization and photocatalytic applications of Zn-doped TiO2 nanoparticles by sol–gel method. *Appl. Nanosci.*, 6, 965–972, 2016.
16. El-Sherbiny, S., Morsy, F., Samir, M., Fouad, O.A., Synthesis, characterization and application of TiO2 nanopowders as special paper coating pigment. *Appl. Nanosci.*, 4, 305–313, 2014.
17. Evcin, A., Arlı, E., Baz, Z., R.Esen, R., Sever, E.G., Characterization of Ag-TiO_2 powders prepared by sol-gel process. *Acta Physica Polonia A*, 132, 608–611, 2017.
18. Birzu, M., Frunza, L., Zgura, I., Cotorobai, V.F., Ganea, C.P., Preda, N., Enculescu, M., Wettability by water contact angle upon the surface of wool fabrics covered with oxide nanoparticles. *Digest J. Nanomater. Biostruct.*, 12, 3, 921–931, 2017.
19. Harikishorea, M., Sandhyarania, M., Venkateswarlua, K., Nellaippana, T.A., Rameshbabua, N., Effect of Ag doping on antibacterial and photocatalytic activity of nanocrystalline TiO_2. *Proc. Mater. Sci.*, 6, 557–566, 2014.
20. Delor, F., Barrois-Oudin, N., Duteurtre, X., Cardinet, C., Lemaire, J., Lacoste, J., Oxidation of rubbers analysed by HATR/IR spectroscopy. *Polym. Degrad. Stab.*, 62, 395–401, 1998.
21. Seentrakoon, B., Junhasavasdikul, B., Chavasiri, W., Enhanced UV-protection and antibacterial properties of natural rubber/rutile-TiO_2 nanocomposites. *Polym. Degrad. Stab.*, 98, 566–578, 2013.

22. Alayande, S., Oluwagbemiga, E., Dare, O., Titus, A.M., Msagati, A., Akinlabi, K., Aiyedun, P.O., Superhydrophobic and superoleophillic surface of porous beaded electrospunpolystrene and polysytrene-zeolite fiber for crude oil-water separation. *Phys. Chem. Earth*, 92, 7–13, 2016.

8

Advanced Technologies for Wastewater Treatment

Asim Ali Yaqoob[1], Claudia Guerrero–Barajas[2], Akil Ahmad[3*], Mohamad Nasir Mohamad Ibrahim[1†] and Mohammed B. Alshammari[3]

[1]*Materials Technology Research Group (MaTRec), School of Chemical Sciences, Universiti Sains Malaysia, Minden, Penang, Malaysia*
[2]*Laboratorio de Biotecnología Ambiental, Departamento de Bioprocesos, Unidad Profesional Interdisciplinaria de Biotecnología, Instituto Politécnico Nacional, Av. Acueducto s/n, Col. Barrio La Laguna Ticomán, Mexico City, Mexico*
[3]*Chemistry Department, College of Science and Humanities, Prince Sattam bin Abdulaziz University, Al-Kharj, Saudi Arabia*

Abstract

Presently, wastewater treatment is a growing and essential topic in water purification. Many factors, such as increasing population, industrialization, urbanization, etc., are majorly responsible for the discharge of different pollutants into the water resources. Previously, many traditional wastewater treatment techniques, such as ion exchange, thermal treatment adsorption, coagulation, electrochemical degradation, and chemical precipitation, were introduced. The mentioned techniques have some drawbacks, such as high-energy demand, production of by-products, toxic effects, cost issues, etc., which make them inefficient for cleaning water. A few important and effective approaches are summarized in this chapter by discussing comprehensive literature based on the wastewater treatment process. Among the abovementioned techniques, the microbial fuel cell (MFC) is considered an effective and effective approach to treating wastewater and has recently attracted researchers and academicians from all over the world. MFCs are a bioelectrochemical approach in which bacterial species serve as biocatalysts for the oxidation of organic matters and reduce the heavy metal content, which leads to the generation of electricity and concurrently cleaning of the wastewater. In this chapter, the role of MFCs in the field of wastewater treatment is briefly summarized.

**Corresponding author*: aj.ahmad@psau.edu.sa
†*Corresponding author*: mnm@usm.my

Shahid-ul-Islam, Abid Hussain Shalla and Mohammad Shahadat (eds.) Green Chemistry for Sustainable Water Purification, (179–202) © 2023 Scrivener Publishing LLC

Finally, the concluding remarks of the literature survey and future recommendations are discussed.

Keywords: Wastewater, wastewater techniques, microbial fuel cell, conventional wastewater treatment, energy generation

8.1 Introduction

Insufficient sources of freshwater availability and high demand for water supply are major issues of the present world due to the growing population and industries. The availability of clean and healthy water is one of the main challenges facing modern society. With urban population growth in developed countries, residual discharge enters into the water bodies due to anthropogenic activities, which produce higher quantities of wastewater [1, 2]. Approximately 15% to 25% of water is drained or drawn from an urban area and the remaining water is recycled to the hydrological system as wastewater [3]. Wastewater is typically a combination of waste and floodwaters in both residential and commercial applications. Industrial wastewater also includes high amounts of chemicals, volatile or metalloid compounds while the pathogenic discharge of domestic wastewater is extremely detrimental [4–6]. As the population increases in many urban areas of the Asian and African continents, wastewater collection is becoming an enormous problem for the municipality. In particular examples, in India, only 24% of household and industrial wastewater is being treated while in Pakistan only 2% of wastewater is treated [7]. Currently, less than 10% of the wastewater is stored in sewage pipes in western African cities and is being treated [8]. Various integrated wastewater storage and treatment facilities have proved to be difficult to maintain in many developed countries. Several areas are supported by decentralized schemes, which are additional durable for long-time service and economic stabilization and compliant with regional sewage demand without any difficulties [9, 10]. In the broad sense, the following factors for water continue to rise increased growth and relocation to drought-prone zones; swift industrialization growth and increased per capita water usage; climate variation in urban areas due to fluctuations in weather restrictions [11–13]. On the other hand, the presence of a huge amount of pathogens and antipathogens in urban and rural water bodies is a threat to water quality [14]. Discharges of wastewater have been recognized as one of the world's problems of water contamination, mainly contributed by municipalities and industrial processing plants. Most of the domestic and agricultural wastewater in many developed worlds is released immediately into water streams without any therapy or primary treatment. Approximately 55% of their waste was dumped without treatment in such a heavily industrialized country as China [13]. The dumping of wastewater without treatment would cause several environmental issues. For example, an excessive amount of heavy metal ions can cause severe health issues as shown in Figure 8.1. As a result, wastewater treatment is needed before it is discharged into natural water sources. The adsorption, biological decay, coagulation, chemical deposition, flocculation, ion exchange, reverse osmosis, and other physical or chemical treatment processes for wastewater treatment have also been reviewed in this chapter. These treatment approaches have some limitations and are not sufficient to treat these toxins. Therefore, in this book chapter, the most innovative and green technology like microbial fuel cells for dealing with wastewater problems is briefly addressed.

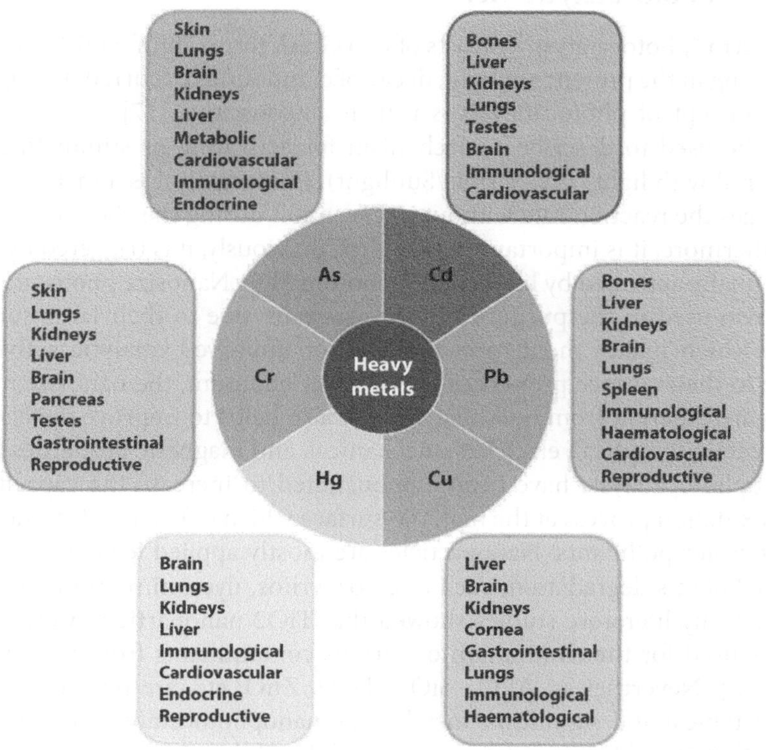

Figure 8.1 Environmental exposure to heavy metals has an effect on the body's major organs and processes (adapted from reference [15] with Elsevier permission).

8.2 Advanced Approaches for Wastewater Treatment

To fulfill the water security requirements for public health, advanced treatment approaches play a vital role in urban and industrial wastewater treatment. Conventional and advanced wastewater management consists of the mixture, settling, suspending, and dissolved solids, organic matter, contaminants, minerals, etc. Nowadays, wastewater recycling technologies are a prime concern for researchers and academicians. The prospects and advances of water reuse have a great impact on the country's economic aspects and open the door for an alternative water source [16]. Furthermore, the use of water regeneration is not a prejudice, as water regeneration and reuse allow more stability in water quality management and thus more possibilities to implement emerging technology. However, the advanced methods, which are used to treat wastewater, are as follows.

8.2.1 Photocatalytic Method

The term "photocatalyst" consists of two Greek terms "film" and "catalytic," meaning in the presence of light, decay of compounds occurred. In science, the concept of photocatalysis is usually unconscious [17]. This method may be used to describe a mechanism for activating or stimulating the material with light (UV/Visible/Sunlight). Photocatalyst is a material that changes the reaction rate without intervention during the chemical phase. Furthermore, it is important to note that, previously, it is triggered by heat and finally activated by light energy photons [18]. Nanosize photocatalysts are also used in the purification of wastewater due to their large surface area which makes them more suitable for improved catalytic activities. Due to their surface properties and distinct quantum, the nanosize materials are different from bulk materials. It also helps to improve the chemical reactivity, optical, electrical, mechanical, and magnetic properties [19]. Nanophotocatalysts have been demonstrated to increase the capacity of the oxidation process at the material surface which efficiently degrades the wastewater pollutants. Nanoparticles are mostly applied for environmental pollutants degradation, such as chlorpyrifos, dyes, nitroaromatics, etc. [20]. Many literature studies showed that TiO2 nanoparticles were effectively used for the elimination of various contaminants from wastewater [21–23]. Nevertheless, Al_2O_3, SiO2, TiO2, ZnO, etc. are reported as the most typical and significant metal oxide nanophotocatalysts for the pollutant's removal from wastewater [24–27]. In addition, nanophotocatalysts may take place heterogeneously or homogenously in two states. Due to the broad variety of water depletion and environmental applications, the most comprehensive analyzed condition is heterogeneous nanophotocatalysis in recent times. In heterogeneity of photocatalyst, an interaction between solid and fluid photocatalyst should be established in advance (such as semiconductor or metal). The term "heterogeneous photocatalysis" is generally used when a light-based photocatalyst interacts with the liquid/gaseous form [28, 29]. The applications of heterogeneous photocatalysis mostly rely on the scaled-up apparatus established on advanced and enhanced solutions. The main thing in reactor design is to enlighten nanocatalysts and optimize the mass transfer, particularly for liquid phases. By using light-emitting diodes and optical fibers, photon transference can be enhanced. But in this region, there are still no productive revolutions found. The positive effect of nanophotocatalyst in the research center for air cleansing and water cleaning has been demonstrated according to the literature. It is also not a perfect means of minimizing the issue at the commercial stage. In the nonexistence of effective photocatalytic structures and reduce photocatalytic skills, the present shortage of substantial industrial implementations is also attributed [19, 22].

The hazardous natural compounds mineralization at 25 °C was demonstrated with a highly accurate and efficient process of detoxification by nanophotocatalysts of water. A highly effective and efficient detoxication procedure was carried out by nanophotocatalysts of water for the hazardous natural compounds mineralization at 25 °C [17]. Overall, the benefits that nanophotocatalysts provide include solidity, low cost, and being biologically friendly. For example, the TiO2 is photostable, but many of these nanophotocatalysts, such as copper nanoparticles, zinc oxides, metal sulfides, etc. have very little natural resilience as a result of photo corrosion occurring [30, 31]. As light shines, they are reduced or oxidized, and their oxidation conditions change when holes or electrons are produced which leads to photocatalyst decomposition and reduces photocatalysts' effectiveness. For this purpose, a nanocomposite needs to be synthesized strongly to produce long-lasting stable photocatalysts. The main benefit of nanosize is the quantum scale, energy bandgap enhancement, and size reduction [32]. In addition, the benefits of photographing as a medium are various, such as being recyclable, low cost, and having good deprivation capacity. Besides all advances, nanophotocatalysts continue to face problems, such as the toxicity and rehabilitation of catalysts. The use and extent of nanophotocatalyst at a higher level is limited due to the abovementioned problems [33]. The scientific community, therefore, concentrates on other nanocomposites for wastewater treatment that can reduce toxicity. Thus, modern photocatalyst, can act in visible scales for the natural

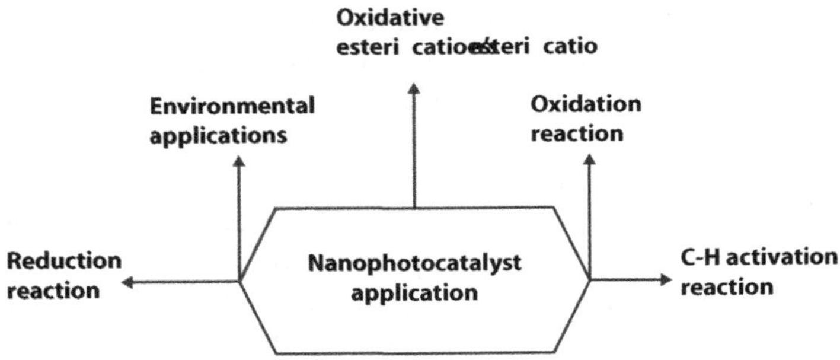

Figure 8.2 Nanophotocatalyst potential applications.

outcome and encourage the use of dialer substances, such as graphene or its derivatives, for the doped photocatalyst to decrease the toxicity effects. Further, the approach to nanophotocatalysts for toxins removal is defined as a highly efficient approach. Figure 8.2 shows several specific uses of nanophotocatalysts.

8.2.1.1 Mechanism of Photocatalysis

The mechanism of photocatalysis was well explained by Yaqoob *et al.* [34]. Figure 8.3 provides a formal overview of the function. The photoelectron

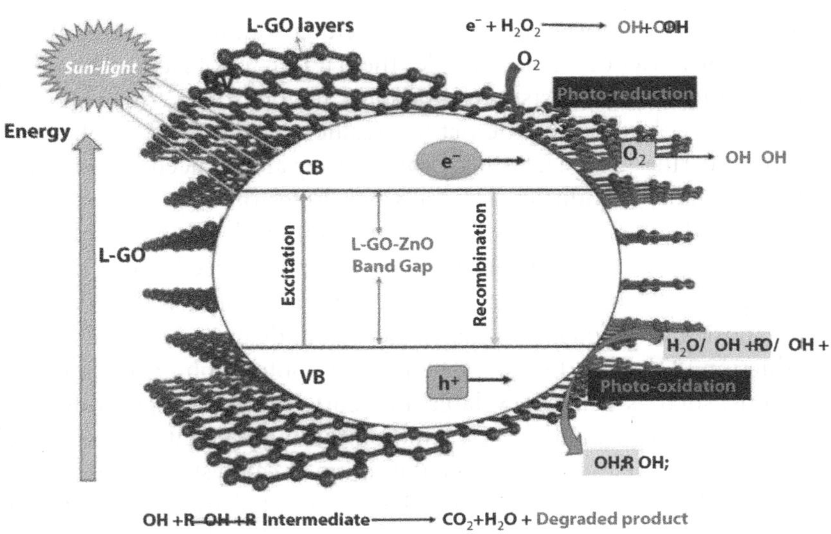

Figure 8.3 A basic photocatalytic mechanism for degradation of pollutants (adapted from reference [34] with springer permission).

is encouraged from the packed valence band in the photocatalyst's blank conduction band (GO-ZnO) by sunlight radiation in the photocatalytic reaction. The absorbed photon is equivalent to or greater than the semiconductor photocatalyst's bandgap, the energy. The valence band (hVB+) hole is caused by sunlight irradiation (excitation). As a result, the following equations show that a set of electrons and hole (e/h +) is formed [35, 36].

$$L - GO - ZnO + h\upsilon \rightarrow GO - ZnO\ e^{-}(CB) + h^{+}(VB) \tag{8.1}$$

$$H_2O + h^{+}(VB)^{\cdot} \rightarrow H^{+} + OH^{\cdot} \tag{8.2}$$

$$e^{-}(CB) + O_2 \rightarrow O_2^{-}\text{(anionic superoxide radical)} \tag{8.3}$$

$$O_2^{-\cdot} + H^{+} \rightarrow HOO^{\cdot} \tag{8.4}$$

$$2HOO^{\cdot} \rightarrow H_2O_2 + O_2 \tag{8.5}$$

$$H_2O_2 \rightarrow 2OH\cdot \tag{8.6}$$

$$OH\cdot + Dye \rightarrow CO_2 + H_2O \tag{8.7}$$

$$e^{-}(CB) + dye \rightarrow Reduction \tag{8.8}$$

$$h^{+}(VB) + dye \rightarrow oxidation \tag{8.9}$$

The OH· is formed in water ionization by the hole produced in the valence band, which serves as an oxidant. The anionic superoxide radical (O_2^-) is supplied with oxygen by the electron of the conduction tape. Following the protonation of superoxide radicals (O_2^-), the formation of peroxide H_2O_2 is produced which breaks further into more reactive hydroxyl radicals (OH·) [37].

8.2.2 Nanomembranes Technology

Nanomembranes are a form of membrane made up of various nanofibers that are used to remove unwanted contaminants from the wastewater. The treatment is carried out at an extreme degree of removal by a dense fouling capacity and has been used as a reverse osmosis technique. Many works have already been reported on membrane technology to create multifunction membranes through different nanosized substances in various polymer-derived membranes [38]. Nanofiltration, reverse osmosis, ultrafiltration, and other water treatment methods were preferably used nano-sized porous membranes. The membrane has a composite layer of porous support. The significant composite sheet is usually carbon-derived (CNT, graphene) which has been fixed into a polymer matrix. This offers exciting and substantial advances in waterborne transport and resistance. For example, CNTs have antimicrobial properties, which can help mitigate fouling, the forming of a biological film, and decrease the likelihood of mechanical disasters [39]. The nanomaterials incorporation into polymeric membranes results in the creation of an expanded membrane with increased fouling resistance and hydrophilicity. Polymer doping is also used to manufacture polymer membranes to avoid bacteria attachment and to prevent biofilm formation at membrane shallow [40, 41]. An antimicrobial agent, such as silver-metal nanoparticles, was used to inactivate bacteria, fungus, etc., and also prevents membrane biofouling [42, 43]. The synthesis and development of thin nanomaterial by doping with metals nanoparticles during surface alteration into an active composite of the thin film. Normally, nanomembrane clogging and membrane fouling have several challenges. Therefore, this issue must be addressed and solved. The prevention from blockage and fouling can be tested by adding superhydrophilic-based nanosize particles to make a thin film nanomembrane. Nanomembrane is often used for wastewater treatment because it has vital characteristics to increase the proliferation of this substance, high unitary, homogeneous, optimal, short-term, readily managed, and produce a high range of reactions [44]. In a nanocomposite membrane, the oxidation of the organic compounds can match those nanophotocatalysts. For example, nanomembranes and films integrated into TiO2 effectively disable and degrade organic contaminants along with blocking the growth of different microorganisms [45]. Nanotechnologies have been developing and increasingly producing many catalytic nanostructured membranes with new properties, such as increased selectivity, high decomposition rates, and higher resistance against fouling [46, 47]. There have been several methods used to synthesize these types of nanomaterials to generate multipurpose functions [17, 48]. Constructive advances in this technique have been achieved through the inclusion in the pore membrane of nanocatalysts including iron-catalyst-dependent free radical and enzymatic catalysis. Oxidative reactions can also be performed without the use of radioactive chemicals to remove toxins and detoxify water. Metallic nanoparticles were immobilized with membranes (e.g., cellulose, chitosan, polysulphone, etc.) for dichlorination and toxic deprivation that include novel properties, such as impediment, high responsiveness, lack of accumulation, and reduction in the surface, have been documented in many tests [49, 50]. In addition, nanotechnology advances yield many important, high discernment, improved absorbency, and high conflict flooding catalytic membranes. Innovative approaches to enhancing the multifunctional characteristics of wastewater management include mixed and bottom-up approaches.

8.2.2.1 Limitations and Future of the Nanomembranes Technology

The key purpose of membranes is to isolate radioactive particles from water supplies in case of drinking water. Filters from nanomembrane are used in water safety measurement [51]. Compared to the filtration technique with conventional approaches, the benefits of nanomembrane are that calcium and magnesium have needed a different ion in the whole phase to compensate for it. The $Na+$ ions are not required for nanomembrane. In comparison with other traditional methods, nanomembrane restrictions typically reduce its performance. The first is nanomembrane fouling, which occurs a few days after using the membrane. This procedure is more costly and ineffective. The nanomembrane has fouling problems since it is completely dependent on operational parameters. The operational parameters are often not suitable, such as high pressure, temperature, and optimization [52]. Membrane stability is the second main constraint. For a long time, the nanomembrane did not maintain equilibrium. After that, its efficiency decreases and does not provide a good result as before. There will also be a need to change the nanomembrane to obtain outstanding performance, but this will lead to many other problems, such as high costs, impurities, etc. [53]. The stability depends on the necessary chemical resistance to material purification. The inconvenience is less reliability, inefficient operating processes, lower selectivity, high maintenance costs, and a reduction in operational efficiency over time. Common drawbacks are therefore not thoroughly investigated. While these nanomembranes have been successfully demonstrated at the laboratory level, they rise to lower prices but are still an obstacle to making them suitable for industrial applications. The commercialization goal needs to be combated through research institutes and manufacturing firms to cross this hurdle to a productive rise. The nanomembrane selectivity should be increased and the nanomembrane resistivity improved to prevent insufficiency of the membrane. The synthesis of new generation membranes may be an important candidate for surface grafting-oriented polymers, such as zwitterionic. However, surface grafting is not reported to be a barrier to internal pore walls [54]. So, suitable antifouling converters become critical to fix on the membrane matrix. The high susceptibility of polyamide membranes to several forms of oxidizers, such as chlorine and ozone is also quite important. Moreover, the manufacture of multifunctional membranes needs considerable attention for better industrial innovation. However, the manufacture of less costly and efficient nanomembrane for water treatment still has many challenges to be addressed.

8.2.3 Utilization of Nanosorbent for Wastewater Treatment

Nanosorbents possess broad functions, such as high-sorption capacity, which make them more suitable for water treatment and more efficient [17]. The nanosorbent is very scarce in trade form, but researchers work hard to manufacture vast quantities of nanosorbents at the trade level. Carbon is the most often notified nanosorbent (e.g., graphite, carbon black, graphene oxide). In addition, there have been also metal/metal oxides and polymer nanosorbents [55–57]. The composite of various materials, such as Ag/carbon, Ag/polyaniline, carbon/TiO2, etc. are of great significance to reduce the impact of toxicity of wastewater. The carbons like CNTs cylinder-shaped nanostructure can appear as single or multiwalled nanotubes. CNTs keep observable sites and maintain sustainable surfaces due to high surface area. To prevent the accumulation of surface sites from decreasing since CNTs have hydrophobic properties on the surface, they must be stabilized [58, 59]. Therefore, the adsorption mechanism is an effective material for the toxins. Polymer nanoadsorbents are also functional for the removal of wastewater toxins like organic and metals [60]. For instance, the use of dendrimer-ultrafiltration has reduced copper ions. They are regenerated easily by pH change and are seen to be biocompatible, biodegrading, and toxic-free. In addition, the percentage of organic dyes or other contaminants removed is reached almost 99% [61]. The zeolites act as an absorbent in which various nanoparticles like silver, copper ions, can be implanted to make it a more significant nanosorbent [62]. It has the advantage of controlling metal quantities and even acts as an antimicrobial agent [63]. In addition, magnetic nanosorbents also play an important part in wastewater remediation and have become a distinctive method to extract various pollutants [64]. A magnetic filtration can also strip certain organic contaminants. Ligands are covered with magnetic nanoparticles at various magnet separation nanosorbents. Many processes, such as magnetic forces, cleaning agents, ion exchange, and so on, have been identified to stimulate these nanosorbents. The redeveloped nanosorbent can be cost-effective and used for further investigation of pollutants to reduce the cost. However, there are certain health hazards to biomass. Studies show that nanoadsorbents, chemical stabilizers, and surface changes are dependent on morphology for the removal of toxic pollutants. To address toxicity problems and health concerns, the focus must also be paid to synthesizing more stable morphology (size and shape). In addition, the adsorbents are characterized by high biodegradability, high biocompatibility, and strong nontoxicity which can be replaced by chemical nanosorbents. Graphene oxide should be suggested to scientists since very emerging nanomaterials can be used as nanosorbents to extract pollutants with their superior properties and receive better results than others [65]. Some nanosorbents and their functions are listed in Table 8.1. Because of their extraordinary physicochemical properties, nanomaterials have a strong interest in explaining ecological problems as well as wastewater decontamination. These characteristics differentiate them from typical sorbents in many areas. To make it perfect for treating the pollutant very quickly, it would have a great surface area, superb adsorption efficiency, fast adsorption times, and balancing intervals. Nanomaterials have a very high absorption rate over some time because of their nanosize. In addition, nanosorbents can be used to separate organic, inorganic contaminants from wastewater. Some issues need to be tackled completely for the commercialization of nanosized decontaminating water sorbents, such as manufacturing scalability, adsorption

Table 8.1 List of nanosorbents and their application in pollutants removal (reproduced from reference [17]).

List of nanosorbents	Functions	References
Polymer fibers	Arsenic and other dangerous metals are removed	[67]
Nanosize-based iron oxides	Removal of hormones and poisonous drugs contaminations	[68]
Nanosize-based metal oxides	Various toxic metals	[69]
Nanoaerogels	Uranium is removed from drinking water	[70]
Regenerable nanosorbents	Organic toxins, sewage water inorganic pollutants	[71]
Graphite oxide	Removal of dyes	[72]
Carbon-derived nanosorbents	Ions of nickel in water	[73]
Nanoclay	Hydrocarbons, Dyes	[74]

calculations, selectivity, substrate consistency, material operating time, etc. [66]. A successful approach to wastewater treatment and the development of new nanosorbent is, however, enormously necessary for the management of wastewater toxic ions and compounds. For the future, the research community works to improve the understanding and processes of adsorption. Researchers, consultants, and experts in the real world are worried about wastewater contaminants with health hazards and they are all committed to finding and clarification the use at the industrial level.

8.2.4 Microbial Fuel Cells as a Sustainable Technique

To remove the different contaminants from water sources, various treatment methods have been studied before [75]. Those procedures are efficient and provide improved performance, but they have few limits e.g., large energy requirements, high costs of service, high chemical utilization, and released toxic waste by-products. The research group has thus introduced a groundbreaking technique for the deprivation of pollutants from wastewater, it is known as Microbial Fuel Cells (MFCs) [76–78]. This theory was first proposed in the early 1900s (1911) by M.C. Potter, using microbes to degrade pollutants and generate energy from bacteria [79].

Later in the 1970s, bacteria were discovered to have redox proteins that are electrochemically active and can be transported by anode electrons [80]. This approach has now attracted considerable exposure than most traditional approaches. MFCs are considered as a green energy generation system that simultaneously reduces wastewater pollutants. The bacteria transfer the electron to the anode and the electron travels via an outer circuit to the cathode chamber, in the presence of an oxidative environment, protons pass directly from anode to cathode [81, 82].

However, there are several factors, like, biocatalyst, internal resistance, chemical concentration, chemical substrates, MFCs modeling, exoelectrogens separated, and electrode characteristics, that play an important role in the efficiency of MFCs [82–84]. The electrode content is a very significant element in making MFCs more commercially viable and prolific. Microbes receive electrons from the anode and also transform the toxic compound into less toxic elements in electrographic electrophs [85, 86]. The MFC technique was established as an optimal technology to eliminate toxins from wastewater through waste-derived catalysts and also generate bioenergy. Centered on their structure, MFCs can be categorized as single, double, and stacked MFCs, sedimental MFCs, benthic MFCs, etc. [87–89].

8.2.4.1 *Mechanism and Application of MFCs in Wastewater Treatment*

The energy production and its mechanism must be known before the metal reduction by bacterial organisms. Bacterial organisms serve as biocatalysts in which the organic matter is oxidized to produce the protons and electrons by the microorganisms species [90]. Some bacterial species of the plant, including *E. coli*, *Clostridium butyric*, *Shewanella* sp., *Rhodoferax ferrireducens*, *Geobacter*, *Bacillus*, *Aeromonas hydrophila*, and *Klebsiella pneumonia* are well recognized as metal reduction genus and exoelectrogens [90–92]. Nevertheless, organic matter is used to oxidize protons and electrons by bacterial organisms. The biochemical reaction is as follows:

$$\text{Anode reaction}: C_6H_{12}O_6 + 6H_2O \longrightarrow 6CO_2 + 24H^+ + 24e^-$$

$$\text{Cathode reaction}: 24H^+ + 24e^- + 6O_2 \longrightarrow 12H_2O$$

$$\text{Overall reaction}: C_6H_{12}O_6 + 6O_2 \longrightarrow 6CO_2 + 6H_2O + \text{Electricity} + \text{Biomass}$$

The biofilm was present on the anode, it transfer the produced electrons and protons. The electrons used the outer way to enter the cathode, while the protons were passed directly by PEM. As seen in Figure 8.4, two main mechanisms were engaged in the electron's transportation from bacteria to the surface of the anode before turning to the cathode. (1) Direct

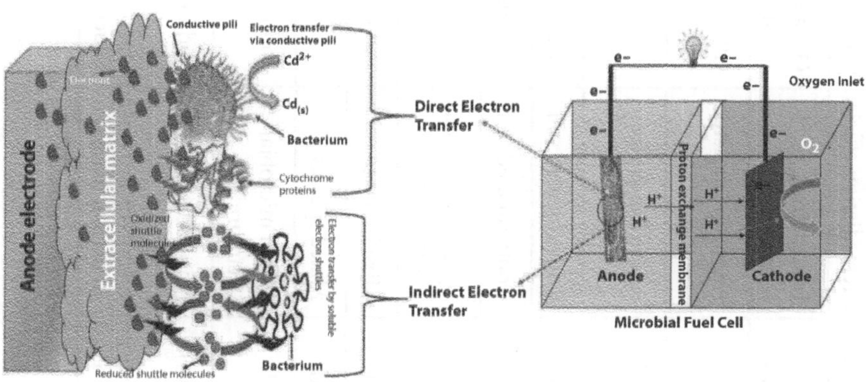

Figure 8.4 MFCs setup with the mechanism of electron transfer from a bacterial cell to the anode electrode. (Copied from reference [94] with Elsevier permission).

Table 8.2 The remediation efficiency of toxic metals through MFCs in previous literature.

Metal ions	Inoculation source	Organic substrate	Anode electrode	Cathode electrode	Removal percentage (%)	Reference
Cr (VI)	Anaerobic digester sludge	Glucose	Graphite felt	Graphite felt	79	[95]
Cr (VI)	Anaerobic sludge	-	Plain carbon cloth	Plain carbon cloth	99	[96]
Cd (II)	Contaminated soil	Sodium acetate	Graphite granules	Carbon felt	31	[97]
Pb (II)	Contaminated soil	Sodium acetate	Graphite granules	Carbon felt	44.1	[97]
V(V)	*Dysgonomonas and Klebsiella*	Acetate	Carbon fiber felt	Carbon fiber felt	60.7	[98]
Cd(II)	Mixed species	Acetate	Graphite felt	Graphite felt	60	[99]
Hg (II)	Mixed species	Acetate	Graphite felt	Graphite felt	55	[99]
Ni (II)	Mixed microbial culture	Sodium acetate	Graphite felt	Graphite felt	95	[100]
Cr (VI)	Anaerobic sludge bed	Sodium acetate	Graphite plate	Graphite plate	82	[100]
Cr (VI)	Anaerobic sludge	Glucose	Graphite brushes	Carbon cloth	99	[101]

(*Continued*)

Table 8.2 The remediation efficiency of toxic metals through MFCs in previous literature. *(Continued)*

Metal ions	Inoculation source	Organic substrate	Anode electrode	Cathode electrode	Removal percentage (%)	Reference
U(VI)	Nuclear waste sludge	Acetate	Graphite felt	Graphite felt	90	[102]
Cu (II)	Soil sludge	Glucose	Carbon felt	Carbon felt	94	[103]
Pb (II)	Wastewater	Sweat potatoes wastes	Graphene oxide	Graphite rod	60.33	[104]
Cd (II)	Wastewater	Sweat potatoes wastes	Graphene oxide/ PANI	Graphite rod	65.51	[104]
Pb (II)	Wastewater	Oil palm trunk sap (OPTs)	L-GO	Graphite rod	85.00	[105]
Pb (II)	Wastewater	OPTs	L-GO-ZnO	Graphite rod	91.07	[105]
Cd (II)	Wastewater	OPTs	L-GO	Graphite rod	83.50	[94]
Cd (II)	Wastewater	OPTs	L-GO-ZnO	Graphite rod	90.00	[94]

electron transfer is subdivided into two methods: electron transportation via conductive pili-type bacterial species or electron transfer via redox-active proteins. Aeromonas hydrophila, Acinetobacter radioresistance strains, Klebsiella pneumonia, and many other bacteria have conductive pili. Pili is a conductive component of the bacterium's body that passes electrons directly to any conductive substance. The bacterial detection method revealed the species that related to the conductive pili-type bacterial, according to the current analysis. (2) The indirect electron transport process employs soluble oxidized/reduced electron shuttles. This method was mostly used by Geobacteraceae and Desulfuromonadaceae to transfer electrons [89, 93]. Another side, the reduction of pollutants was carried out. There are many studies on pollutant reduction via MFCs, especially metal, some are summarized in Table 8.2. For example, there was also the Cd (II) to Cd reduction process. As a result of the reduction, Cd (II) to Cd(s) ensures that the soluble metal ion condition is transformed into a solid state of insolubility. Compared to the soluble Cd condition, it is less poisonous. The reduction from Cd (II) to Cd is as follows [94]:

$$Cd^{2+} + 2e^- \longrightarrow Cd_{(s)}$$

$$2Cd^{2+} + 2H_2O \longrightarrow 2CdO + 4H^+$$

$$CdO + 2e^- + 2H^+ \longrightarrow Cd_{(s)} + H_2O$$

8.3 Conclusion and Future Recommendations

The treatment of wastewater is closely linked with the effluent quality requirements and prospects. The wastewater treatment processes are constructed in such a way as to improve water quality. Thus, the choice of techniques best suited to one particular plant should consider various factors, such as quantity and quality of wastewater, accessibility of land, economic flexibility, etc. Besides numerous technological advances, 100% treatment has yet to be done to confirm improved ecological protection. Water management is a critical ecological protection mechanism to be promoted in the developed world. This book chapter reviewed the application of various adsorbents, such as nanophotocatalyst, nanosorbents, nanomembrane, and MFCs technologies, for wastewater treatment. At the moment, these are well-known modern technologies but all the MFCs attract more attention due to their eco-friendly nature. For the treatment of wastewater, developing a suitable treatment technology should prime the work of researchers and academicians and ensured the best water quality is achieved. In addition, improvements in public education are proposed to ensure knowledge of the technology and its environmental and economic advantages.

Acknowledgments

This chapter was supported by Universiti Sains Malaysia (Malaysia).

References

1. Bhaduri, A., Bogardi, J., Siddiqi, A., Voigt, H., Vörösmarty, C., Pahl-Wostl, C., Bunn, S.E., Shrivastava, P., Lawford, R., Foster, S., Achieving sustainable development goals from a water perspective. *Front. Environ. Sci.*, 4, 64, 2016.
2. Li, X., Song, J., Lin, T., Dixon, J., Zhang, G., Ye, H., Urbanization and health in china, thinking of the national, local and individual levels. *Environ. Health*, 15, 113–123, 2016.
3. Qadir, M., Wichelns, D., Raschid-Sally, L., McCornick, P.G., Drechsel, P., Bahri, A., Minhas, P., The challenges of wastewater irrigation in developing countries. *Agric. Water Manage.*, 97, 561–568, 2010.
4. Cui, J., Zheng, L., Deng, Y., Emergency water treatment with ferrate (VI) in response to natural disasters. *Environ. Sci.: Water Res. Technol.*, 4, 359–368, 2018.
5. Sonne, A. and Ghate, R., Developments in wastewater treatment methods. *Desalination*, 167, 55–63, 2004.
6. Shukla, S., Kumar, S., Gita, S., Bharti, V., Kumar, K., Bhuvaneswari, G.R., Recent technologies for wastewater treatment: A brief review, in: *Wastewater Management through Aquaculture*, B.B. Jana, R.N. Mandal, P. Jayasankar (Eds.), pp. 225–234, Springer Nature, Singapore, 2018.
7. Minhas, P. and Samra, J., Quality assessment of water resources in the indo-gangetic basin part in India. *Tech. Bull. No. 1/2003*, Central Soil Salinity Research Institute, Karnal-132001, India, p. 68, 2003.
8. Drechsel, P., Graefe, S., Sonou, M., Cofie, O.O., Informal irrigation in urban West Africa: An overview, IWMI Research Reports H039249, International Water Management Institute, 2006.
9. Raschid-Sally, L. and Parkinson, J., Wastewater reuse for agriculture and aquaculture-current and future perspectives for low-income countries. *Waterlines*, 23, 2–4, 2004.
10. Crini, G. and Lichtfouse, E., Advantages and disadvantages of techniques used for wastewater treatment. *Environ. Chem. Lett.*, 17, 145–155, 2019.

11. Dhote, J., Ingole, S., Chavhan, A., Review on wastewater treatment technologies. *Int. J. Eng. Res. Technol.*, 1, 1–10, 2012.
12. Pearce, G., UF/MF pre-treatment to RO in seawater and wastewater reuse applications: a comparison of energy costs. *Desalination*, 222, 66–73, 2008.
13. Seow, T.W., Lim, C.K., Nor, M.H.M., Mubarak, M., Lam, C.Y., Yahya, A., Ibrahim, Z., Review on wastewater treatment technologies. *Int. J. Appl. Environ. Sci.*, 11, 111–126, 2016.
14. Minhas, P., Sharma, N., Yadav, R., Joshi, P., Prevalence and control of pathogenic contamination in some sewage irrigated vegetable, forage and cereal grain crops. *Biores. Technol.*, 97, 1174–1178, 2006.
15. García-Niño, W.R. and Pedraza-Chaverrí, J., Protective effect of curcumin against heavy metals-induced liver damage. *Food Chem. Toxicol.*, 69, 182–201, 2014.
16. Yaqoob, A.A., Ibrahim, M.N.M., Umar, K., Biomass-derived composite anode electrode: Synthesis, characterizations, and application in microbial fuel cells (MFCs). *J. Environ. Chem. Eng.*, 9, 106111, 2021.
17. Yaqoob, A.A., Parveen, T., Umar, K., Mohamad Ibrahim, M.N., Role of nanomaterials in the treatment of wastewater: A review. *Water*, 12, 495, 2020.
18. Gomes, J., Lincho, J., Domingues, E., Quinta-Ferreira, R.M., Martins, R.C., N–TiO2 photocatalysts: A review of their characteristics and capacity for emerging contaminants removal. *Water*, 11, 373, 2019.
19. Umar, K., Aris, A., Parveen, T., Jaafar, J., Majid, Z.A., Reddy, A.V.B., Talib, J., Synthesis, characterization of Mo and Mn doped ZnO and their photocatalytic activity for the decolorization of two different chromophoric dyes. *Appl. Catal. A: Gen.*, 505, 507–514, 2015.
20. Samanta, H., Das, R., Bhattachajee, C., Influence of nanoparticles for wastewater treatment-a short review. *Austin Chem. Eng.*, 3, 1036, 2016.
21. Umar, K., Parveen, T., Khan, M.A., Ibrahim, M.N.M., Ahmad, A., Rafatullah, M., Degradation of organic pollutants using metal-doped TiO2 photocatalysts under visible light: A comparative study. *Desal. Water Treat*, 161, 275–282, 2019.
22. Umar, K., Ibrahim, M.N.M., Ahmad, A., Rafatullah, M., Synthesis of Mn-doped TiO 2 by novel route and photocatalytic mineralization/intermediate studies of organic pollutants. *Res. Chem. Intermed.*, 45, 2927–2945, 2019.
23. Haque, M., Khan, A., Umar, K., Mir, N.A., Muneer, M., Harada, T., Matsumura, M., Synthesis, characterization and photocatalytic activity of visible light induced Ni-doped TiO2. *Energy Environ. Focus*, 2, 73–78, 2013.
24. Yaqoob, A.A., Serrà, A., Ibrahim, M.N.M., Advances and challenges in developing efficient graphene oxide-based ZnO photocatalysts for dye photooxidation. *Nanomaterials*, 10, 932, 2020.
25. Umar, K., Water contamination by organic-pollutants: TiO 2 photocatalysis, in: *Modern Age Environmental Problems and their Remediation*, pp. 95–109, Springer, Switzerland AG., 2018.

26. Umar, K., Aris, A., Ahmad, H., Parveen, T., Jaafar, J., Majid, Z.A., Reddy, A.V.B., Talib, J., Synthesis of visible light active doped TiO 2 for the degradation of organic pollutants—Methylene blue and glyphosate. *J. Anal. Sci. Technol.*, 7, 1–8, 2016.
27. Sultana, S., Khan, M.Z., Umar, K., Ahmed, A.S., Shahadat, M., SnO2–SrO based nanocomposites and their photocatalytic activity for the treatment of organic pollutants. *J. Mol. Struct.*, 1098, 393–399, 2015.
28. Kohtani, S., Kawashima, A., Miyabe, H., Stereoselective organic reactions in heterogeneous semiconductor photocatalysis. *Front. Chem.*, 7, 630, 2019.
29. Ahmad, A., Lokhat, D., Wang, Y., Rafatullah, M., Recent advances in nanofiltration membrane techniques for separation of toxic metals from wastewater, in: *Nanotechnology for Sustainable Water Resources*, A.K. Mishra, C.M. Hussain (Eds.), pp. 477–500, Wiley: Scrivener Publishing LLC, Beverly, MA, 2018.
30. Umar, K., Haque, M., Muneer, M., Harada, T., Matsumura, M., Mo, Mn and La doped TiO2: synthesis, characterization and photocatalytic activity for the decolourization of three different chromophoric dyes. *J. Alloy Compounds*, 578, 431–438, 2013.
31. Mir, N.A., Khan, A., Umar, K., Muneer, M., Photocatalytic study of a xanthene dye derivative, phloxine B in aqueous suspension of TiO2: Adsorption isotherm and decolourization kinetics. *Energy Environ. Focus*, 2, 208–216, 2013.
32. Rajabi, H.R., Shahrezaei, F., Farsi, M., Zinc sulfide quantum dots as powerful and efficient nanophotocatalysts for the removal of industrial pollutant. *J. Mater. Sci.: Mater. Electron.*, 27, 9297–9305, 2016.
33. Mahmoodi, N.M. and Arami, M., Degradation and toxicity reduction of textile wastewater using immobilized titania nanophotocatalysis. *J. Photochem. Photobiol. B: Biol.*, 94, 20–24, 2009.
34. Yaqoob, A.A., Umar, K., Adnan, R., Ibrahim, M.N.M., Rashid, M., Graphene oxide–ZnO nanocomposite: An efficient visible light photocatalyst for degradation of rhodamine B. *Appl. Nanosci.*, 11, 1291–1302, 2021.
35. Ansari, S.A., Ansari, S., Foaud, H., Cho, M.H., Facile and sustainable synthesis of carbon-doped ZnO nanostructures towards the superior visible light photocatalytic performance. *New J. Chem.*, 41, 9314–9320, 2017.
36. Peter, C., Anku, W.W., Sharma, R., Joshi, G.M., Shukla, S.K., Govender, P.P., N-doped ZnO/graphene oxide: A photostable photocatalyst for improved mineralization and photodegradation of organic dye under visible light. *Ionics*, 25, 327–339, 2019.
37. Viswanathan, B., Photocatalytic degradation of dyes: An overview. *Curr. Catal.*, 7, 99–121, 2018.
38. Umar, K., Yaqoob, A.A., Ibrahim, M., Parveen, T., Safian, M., Environmental applications of smart polymer composites. *Smart Polym. Nanocompos. Biomed. Environ. Appl.*, 15, 295–320, 2020.

39. Hogen-Esch, T., Pirbazari, M., Ravindran, V., Yurdacan, H.M., Kim, W., High performance membranes for water reclamation using polymeric and nanomaterials. *US Patents,* Sr. No. US, 10, 456, 754 B2, 2019.
40. Kochkodan, V. and Hilal, N., A comprehensive review on surface modified polymer membranes for biofouling mitigation. *Desalination*, 356, 187–207, 2015.
41. Saleh, T.A., Parthasarathy, P., Irfan, M., Advanced functional polymer nanocomposites and their use in water ultra-purification. *Trends Environ. Anal. Chem.*, 24, e00067, 2019.
42. Yaqoob, A.A., Umar, K., Ibrahim, M.N.M., Silver nanoparticles: various methods of synthesis, size affecting factors and their potential applications–A review. *Appl. Nanosci.*, 10, 1369–1378, 2020.
43. Yaqoob, A.A., Ahmad, H., Parveen, T., Ahmad, A., Oves, M., Ismail, I.M., Qari, H.A., Umar, K., Mohamad Ibrahim, M.N., Recent advances in metal decorated nanomaterials and their various biological applications: A review. *Front. Chem.*, 8, 341, 2020.
44. Hirata, K., Watanabe, H., Kubo, W., Nanomembranes as a substrate for ultrathin lightweight devices. *Thin Solid Films*, 676, 8–11, 2019.
45. Gopalakrishnan, I., Samuel, R.S., Sridharan, K., Nanomaterials-based adsorbents for water and wastewater treatments, in: *Emerging Trends of Nanotechnology in Environment and Sustainability*, K. Sridharan (Ed.), pp. 89–98, Springer, Switzerland AG., 2018.
46. Liu, S., Wang, Y., Zhou, X., Han, W., Li, J., Sun, X., Shen, J., Wang, L., Improved degradation of the aqueous flutriafol using a nanostructure macroporous PbO2 as reactive electrochemical membrane. *Electrochim. Acta*, 253, 357–367, 2017.
47. Shetti, N.P., Bukkitgar, S.D., Reddy, K.R., Reddy, C.V., Aminabhavi, T.M., Nanostructured titanium oxide hybrids-based electrochemical biosensors for healthcare applications. *Coll. Surf. B: Biointerfaces*, 178, 385–394, 2019.
48. Daud, N.N., Ahmad, A., Yaqoob, A.A., Ibrahim, M.N.M., Application of rotten rice as a substrate for bacterial species to generate energy and the removal of toxic metals from wastewater through microbial fuel cells. *Environ. Sci. Pollut. Res.*, 2, 1–2, 2021.
49. Mekaru, H., Lu, J., Tamanoi, F., Development of mesoporous silica-based nanoparticles with controlled release capability for cancer therapy. *Adv. Drug Deliv. Rev.*, 95, 40–49, 2015.
50. Jawed, A., Saxena, V., Pandey, L.M., Engineered nanomaterials and their surface functionalization for the removal of heavy metals: A review. *J. Water Process Eng.*, 33, 101009, 2020.
51. Jawaid, M. and Khan, M.M., *Polymer-based nanocomposites for energy and environmental applications*, pp. 1–679, Woodhead Publishing, Cambridge, 2018.
52. Rashidi, H.R., Sulaiman, N.M.N., Hashim, N.A., Hassan, C.R.C., Ramli, M.R., Synthetic reactive dye wastewater treatment by using nano-membrane filtration. *Desalin. Water Treat*, 55, 86–95, 2015.

53. Belloň, T., Polezhaev, P., Vobecká, L., Slouka, Z., Fouling of a heterogeneous anion-exchange membrane and single anion-exchange resin particle by ssdna manifests differently. *J. Membr. Sci.*, 572, 619–631, 2019.
54. Ying, Y., Ying, W., Li, Q., Meng, D., Ren, G., Yan, R., Peng, X., Recent advances of nanomaterial-based membrane for water purification. *Appl. Mater. Today*, 7, 144–158, 2017.
55. Yu, L., Ruan, S., Xu, X., Zou, R., Hu, J., One-dimensional nanomaterial-assembled macroscopic membranes for water treatment. *Nano Today*, 17, 79–95, 2017.
56. Ahmad, A., Mohd-Setapar, S.H., Chuong, C.S., Khatoon, A., Wani, W.A., Kumar, R., Rafatullah, M., Recent advances in new generation dye removal technologies: Novel search for approaches to reprocess wastewater. *RSC Adv.*, 5, 30801–30818, 2015.
57. Hussain, I., Hussain, A., Ahmad, A., Rahman, H., Alajmi, M.F., Ahmed, F., Amir, S., New generation graphene oxide for removal of polycyclic aromatic hydrocarbons, in: *Graphene-Based Nanotechnologies for Energy and Environmental Applications*, M. Jawaid, A. Ahmad, D. Lokhat (Eds.), pp. 241–266, Elsevier, Woodhead Publishing, Cambridge, 2019.
58. Ganesh, E., Single walled and multi walled carbon nanotube structure, synthesis and applications. *Int. J. Innov. Technol. Exp. Eng.*, 2, 311–320, 2013.
59. Kim, K.S., Cota-Sanchez, G., Kingston, C.T., Imris, M., Simard, B., Soucy, G., Large-scale production of single-walled carbon nanotubes by induction thermal plasma. *J. Phys. D: Appl. Phys.*, 40, 2375, 2007.
60. Fuwad, A., Ryu, H., Malmstadt, N., Kim, S.M., Jeon, T.-J., Biomimetic membranes as potential tools for water purification: Preceding and future avenues. *Desalination*, 458, 97–115, 2019.
61. Peng, F., Xu, T., Wu, F., Ma, C., Liu, Y., Li, J., Zhao, B., Mao, C., Novel biomimetic enzyme for sensitive detection of superoxide anions. *Talanta*, 174, 82–91, 2017.
62. Giwa, A., Hasan, S., Yousuf, A., Chakraborty, S., Johnson, D., Hilal, N., Biomimetic membranes: A critical review of recent progress. *Desalination*, 420, 403–424, 2017.
63. Asyikin binti Abdul Aziz, Z., Ahmad, A., Hamidah Mohd-Setapar, S., Hassan, H., Lokhat, D., Amjad Kamal, M., Recent advances in drug delivery of polymeric nano-micelles. *Curr. Drug Metabol.*, 18, 16–29, 2017.
64. Sahebi, S., Sheikhi, M., Ramavandi, B., A new biomimetic aquaporin thin film composite membrane for forward osmosis: Characterization and performance assessment. *Desalin. Water Treat*, 148, 42–50, 2019.
65. Vunain, E., Mishra, A., Mamba, B., Dendrimers, mesoporous silicas and chitosan-based nanosorbents for the removal of heavy-metal ions: A review. *Int. J. Biol. Macromol.*, 86, 570–586, 2016.
66. Khajeh, M., Laurent, S., Dastafkan, K., Nanoadsorbents: Classification, preparation, and applications (with emphasis on aqueous media). *Chem. Rev.*, 113, 7728–7768, 2013.

67. Yadav, V.B., Gadi, R., Kalra, S., Clay based nanocomposites for removal of heavy metals from water: A review. *J. Environ. Manage.*, 232, 803–817, 2019.
68. Unuabonah, E.I. and Taubert, A., Clay–polymer nanocomposites (CPNs): Adsorbents of the future for water treatment. *Appl. Clay Sci.*, 99, 83–92, 2014.
69. Wang, Y., Zhang, Y., Hou, C., Liu, M., Mussel-inspired synthesis of magnetic polydopamine–chitosan nanoparticles as biosorbent for dyes and metals removal. *J. Taiwan Inst. Chem. Eng.*, 61, 292–298, 2016.
70. Krstić, V., Urošević, T., Pešovski, B., A review on adsorbents for treatment of water and wastewaters containing copper ions. *Chem. Eng. Sci.*, 192, 273–287, 2018.
71. Lee, X., Lee, L., Foo, L., Tan, K., Hassell, D., Evaluation of carbon-based nanosorbents synthesised by ethylene decomposition on stainless steel substrates as potential sequestrating materials for nickel ions in aqueous solution. *J. Environ. Sci.*, 24, 1559–1568, 2012.
72. Sun, X., Liu, Z., Zhang, G., Qiu, G., Zhong, N., Wu, L., Cai, D., Wu, Z., Reducing the pollution risk of pesticide using nano networks induced by irradiation and hydrothermal treatment. *J. Environ. Sci. Health Part B*, 50, 901–907, 2015.
73. Rodovalho, F.L., Capistrano, G., Gomes, J.A., Sodré, F.F., Chaker, J.A., Campos, A.F., Bakuzis, A.F., Sousa, M.H., Elaboration of magneto-thermally recyclable nanosorbents for remote removal of toluene in contaminated water using magnetic hyperthermia. *Chem. Eng. J.*, 302, 725–732, 2016.
74. Kyzas, G.Z. and Matis, K.A., Nanoadsorbents for pollutants removal: A review. *J. Mol. Liquids*, 203, 159–168, 2015.
75. Yaqoob, A.A., Mohamad Ibrahim, M.N., Rafatullah, M., Chua, Y.S., Ahmad, A., Umar, K., Recent advances in anodes for microbial fuel cells: An overview. *Materials*, 13, 2078, 2020.
76. Logan, B.E., Scaling up microbial fuel cells and other bioelectrochemical systems. *Appl. Microbial. Biotechnol.*, 85, 1665–1671, 2010.
77. Kim, J.R., Min, B., Logan, B.E., Evaluation of procedures to acclimate a microbial fuel cell for electricity production. *Appl. Microbiol. Biotechnol.*, 68, 23–30, 2005.
78. Logan, B.E., Wallack, M.J., Kim, K.-Y., He, W., Feng, Y., Saikaly, P.E., Assessment of microbial fuel cell configurations and power densities. *Environ. Sci. Technol. Lett.*, 2, 206–214, 2015.
79. Rahimnejad, M., Adhami, A., Darvari, S., Zirepour, A., Oh, S.-E., Microbial fuel cell as new technology for bioelectricity generation: A review. *Alexandria Eng. J.*, 54, 745–756, 2015.
80. Drendel, G., Mathews, E.R., Semenec, L., Franks, A.E., Microbial fuel cells, related technologies, and their applications. *Appl. Sci.*, 8, 2384, 2018.
81. Yaqoob, A.A., Khatoon, A., Mohd Setapar, S.H., Umar, K., Parveen, T., Mohamad Ibrahim, M.N., Ahmad, A., Rafatullah, M., Outlook on the role of microbial fuel cells in remediation of environmental pollutants with electricity generation. *Catalysts*, 10, 819, 2020.

82. Yaqoob, A.A., Ibrahim, M.N.M., Rodríguez-Couto, S., Development and modification of materials to build cost-effective anodes for microbial fuel cells (MFCs): An overview. *Biochem. J.*, 164, 107779, 2020.
83. Araneda, I., Tapia, N.F., Lizama Allende, K., Vargas, I.T., Constructed wetland-microbial fuel cells for sustainable greywater treatment. *Water*, 10, 940, 2018.
84. Ahmad, A., Khatoon, A., Umar, M.F., Abbas, S.Z., Rafatullah, M., Nanocomposite materials as electrode materials in microbial fuel cells for the removal of waterp, in: *Emerging Carbon-Based Nanocomposite for Environmental Application*, A.K. Mishra, C.M. Hussain (Eds.), pp. 213–235, John Wiley and Sons Inc. New York, USA, 2020.
85. Palanisamy, G., Jung, H.-Y., Sadhasivam, T., Kurkuri, M.D., Kim, S.C., Roh, S.-H., A comprehensive review on microbial fuel cell technologies: Processes, utilization, and advanced developments in electrodes and membranes. *J. Clean. Prod.*, 221, 598–621, 2019.
86. Yaqoob, A.A., Ibrahim, M.N.M., Guerrero-Barajas, C., Modern trend of anodes in microbial fuel cells (MFCs): An overview. *Environ. Technol. Innov.*, 23, 101579, 2021.
87. Umar, M.F., Rafatullah, M., Abbas, S.Z., Mohamad Ibrahim, M.N., Ismail, N., Advancement in benthic microbial fuel cells toward sustainable bioremediation and renewable energy production. *Int. J. Environ. Res. Public Health*, 18, 3811, 2021.
88. Fadzli, F.S., Rashid, M., Yaqoob, A.A., Ibrahim, M.N.M., Electricity generation and heavy metal remediation by utilizing yam (Dioscorea alata) waste in benthic microbial fuel cells (BMFCs). *Biochem. Eng. J.*, 172, 108067, 2021.
89. Abbas, S., Rafatullah, M., Hossain, K., Ismail, N., Tajarudin, H., Khalil, H.A., A review on mechanism and future perspectives of cadmium-resistant bacteria. *Int. J. Environ. Sci. Technol.*, 15, 243–262, 2018.
90. Asim, A.Y., Mohamad, N., Khalid, U., Tabassum, P., Akil, A., Lokhat, D., Siti, H., A glimpse into the microbial fuel cells for wastewater treatment with energy generation. *Desalination Water Treat.*, 214, 379–389, 2021.
91. Umar, M.F., Abbas, S.Z., Mohamad Ibrahim, M.N., Ismail, N., Rafatullah, M., Insights into advancements and electrons transfer mechanisms of electrogens in benthic microbial fuel cells. *Membranes*, 10, 205, 2020.
92. Chuo, S.C., Mohamed, S.F., Mohd Setapar, S.H., Ahmad, A., Jawaid, M., Wani, W.A., Yaqoob, A.A., Mohamad Ibrahim, M.N., Insights into the current trends in the utilization of bacteria for microbially induced calcium carbonate precipitation. *Materials*, 13, 4993, 2020.
93. Abbas, S.Z., Rafatullah, M., Ismail, N., Syakir, M.I., A review on sediment microbial fuel cells as a new source of sustainable energy and heavy metal remediation: mechanisms and future prospective. *Int. J. Energy Res.*, 41, 1242–1264, 2017.
94. Yaqoob, A.A., Serrà, A., Ibrahim, M.N.M., Yaakop, A.S., Self-assembled oil palm biomass-derived modified graphene oxide anode: An efficient medium

for energy transportation and bioremediating Cd (II) via microbial fuel cells. *Arabian J. Chem.*, 14, 103121, 2021.

95. Wu, X., Zhu, X., Song, T., Zhang, L., Jia, H., Wei, P., Effect of acclimatization on hexavalent chromium reduction in a biocathode microbial fuel cell. *Biores. Technol.*, 180, 185–191, 2015.
96. Gangadharan, P. and Nambi, I.M., Hexavalent chromium reduction and energy recovery by using dual-chambered microbial fuel cell. *Water Sci. Technol.*, 71, 353–358, 2015.
97. Habibul, N., Hu, Y., Sheng, G.-P., Microbial fuel cell driving electrokinetic remediation of toxic metal contaminated soils. *J. Hazard. Mater.*, 318, 9–14, 2016.
98. Qiu, R., Zhang, B., Li, J., Lv, Q., Wang, S., Gu, Q., Enhanced vanadium (V) reduction and bioelectricity generation in microbial fuel cells with biocathode. *J. Power Sources*, 359, 379–383, 2017.
99. Gai, R., Liu, Y., Liu, J., Yan, C., Jiao, Y., Cai, L., Zhang, L., Behavior of copper, nickel, cadmium and mercury ions in anode chamber of microbial fuel cells. *Int. J. Electrochem. Sci.*, 13, 3050–3062, 2018.
100. Li, F., Jin, C., Choi, C., Lim, B., Simultaneous removal and/or recovery of cr(VI) a (III) using a double mfc technique. *Environ. Eng. Manage. J. (EEMJ)*, 18, 235–242, 2019.
101. Wang, H., Song, X., Zhang, H., Tan, P., Kong, F., Removal of hexavalent chromium in dual-chamber microbial fuel cells separated by different ion exchange membranes. *J. Hazard. Mater.*, 384, 121459, 2020.
102. Vijay, A., Khandelwal, A., Chhabra, M., Vincent, T., Microbial fuel cell for simultaneous removal of uranium (VI) and nitrate. *Chem. Eng. J.*, 388, 124157, 2020.
103. Zhang, J., Cao, X., Wang, H., Long, X., Li, X., Simultaneous enhancement of heavy metal removal and electricity generation in soil microbial fuel cell. *Ecotoxicol. Environ. Saf.*, 192, 110314, 2020.
104. Yaqoob, A.A., Mohamad Ibrahim, M.N., Umar, K., Bhawani, S.A., Khan, A., Asiri, A.M., Khan, M.R., Azam, M., AlAmmari, A.M., Cellulose derived graphene/polyaniline nanocomposite anode for energy generation and bioremediation of toxic metals via benthic microbial fuel cells. *Polymers*, 13, 135, 2021.
105. Yaqoob, A.A., Ibrahim, M.N.M., Yaakop, A.S., Umar, K., Ahmad, A., Modified graphene oxide anode: A bioinspired waste material for bioremediation of Pb2+ with energy generation through microbial fuel cells. *Chem. Eng. J.*, 417, 128052, 2020.

9

PDMS-Supported Composite Materials as Oil Absorbent

Nur Anis Syazmin, Mohammad Shahadat*, Mohd Rizal Razali and Rohana Adnan

School of Chemical Sciences, Universiti Sains Malaysia, USM Penang, Malaysia

Abstract
The problem of effective oil-water separation has prompted a rise in fundamental research in recent years, which has had a considerable influence on the development of practical applications. Owing to their remarkable absorption properties, chemical stability, excellent selectivity, and good recyclability, superhydrophobic and super oleophilic nature show enormous potential as absorbents to clean up an oil spill. The most current manufacturing processes for polydimethylsiloxane (PDMS), the most often used oil-absorbent sponge to separate oil-water is discussed in this chapter. A study of the various manufacturing methods will be beneficial for individuals who are new to the matter since it will offer a brief knowledge of the possible uses of a PDMS sponge manufactured using a particular approach. Future selective oil absorbents with significant economic potential will need a thorough understanding of how the manufacturing process, structure, and performance interact.

***Keywords*:** Oil-water separation, polydimethylsiloxane (PDMS), manufacture of PDMS sponges

9.1 Introduction

Oil spills and oily industrial wastewater treatment have grown more common as a result of the fast expansion of the petrochemical and marine sectors. This has posed a severe issue across the nation. Oil spills and organic

**Corresponding author*: mdshahadat93@gmail.com

Shahid-ul-Islam, Abid Hussain Shalla and Mohammad Shahadat (eds.) Green Chemistry for Sustainable Water Purification, (203–222) © 2023 Scrivener Publishing LLC

liquid spills have impacted negatively on natural systems in recent years, with major health, social, and economic ramifications for the people who are affected [1]. For example, in April 2010, an explosion on a drilling site off the coast of Louisiana led to a tremendous mass of crude oil being released into the Gulf of Mexico, resulting in a long-term effect on the environment [2]. There might be a vast variety of environmental consequences from the oil spill in the water, including physical harm to animals and species, habitat destruction, and ecological disturbance [3]. It has resulted in the development of a broad range of technologies for cleaning oil from water. It may be divided into the following categories: Spillages are collected off the ocean's surface, mixed with water with the use of emulsifiers to accelerate natural decomposition and then burnt on the place to prevent further pollution [4]. On the other hand, natural degradation results in material waste, while *in situ* burning causes energy waste [5].

One of the most practical solutions available is to develop a lightweight porous absorbent with a maximum oil uptake capacity [1]. In terms of having strong hydrophobicity and oleophilic, a great oil spill cleanup sorbent material also has a relatively high sorption potential, low water pickup, and is very affordable and simple to utilize [6]. A good absorbent's surface has both microscale and nanoscale roughness, and air pockets may be used to effectively discard a water droplet from its surface. Super-hydrophobic surfaces with water contact angles (WCA) more than 150° and water slide angles (WSA) less than 10° are generally referred to as superhydrophobic surfaces because they resist water extremely efficiently. These materials might potentially be helpful as oil absorbents [7]. Therefore, developing porous materials with variable relative permeability along with low surface energy is a viable research area for commercial oil separation [8].

Polydimethylsiloxane (PDMS) has a high hydrophobicity combined with flexibility and oleophilic, as well as thermal, mechanical, and chemical durability, and it is also simple to manufacture [1]. PDMS is a flexible polymer with good physical and chemical characteristics, in addition to its structural elements (Si-O-Si angles; Si-O bond length, dissociation energy, and freedom of rotation; weak intermolecular forces) [9]. Aside from that, due to their hydrophobic nature, PDMS sponges have the potential to gather oils and organic solvents from water on their own without the assistance of other materials [10]. Considering the high chain mobility of the material, even though PDMS has a low glass transition temperature (120°C), introducing small holes or surface functional groups into the polymer may result in a quick restoration of the original structure [8]. The present chapter provides an overview of the general characteristics of polydimethylsiloxane (PDMS), followed by a description of how to fabricate PDMS sponges for use as oil/water separation materials using typical synthetic procedures.

9.2 Fabrications Techniques of PDMS Sponges as Oil Absorbent

To develop PDMS sponges as an oil absorbent various methods have been proposed: using sacrificial templates, such as sugar [4], salt [11] and nickel foam [12], emulsion templating [13], phase separation [14], 3D printing method [15], and gas foaming method [16]. However, the characteristics and performance of PDMS depend on the fabrication strategies where they determine the bulk and surface structure. PDMS is also highly superhydrophobic due to the precise incorporation of micropores and surface roughness [8].

9.2.1 Sacrificial Templates

To create PDMS sponges that can be disintegrated or removed selectively, porogen, also known as a solid template group, was utilized. PDMS skeleton with linked voids was revealed when the leachate particles were either dissolved or removed from sample [17]. The most often used sacrificial templates are salt or sugar cubes, and their production is simple since no complex laboratory equipment or toxic solvents are necessary [17]. The PDMS crosslink in this procedure when the PDMS precursors are combined with the scarifying template particles [8]. Pores are generated as a result of the shape of the PDMS template particles after they have been removed via particulate leaching with salt or sugar cube [8].

It is a common practice to use water for the leaching process. If it is essential, the extractable salt or sugar particles/cubes can be retrieved after they have been dispersed in the water. As a consequence, sugar leaching technologies are being developed for use in large-scale industrial applications as well. This method does not need complicated equipment, and the leachate may be retrieved without any disturbance. Choi *et al.*, for example, created PDMS sponges using sugar as a sacrificial template. The template's absorption performance can be improved by adequately mixing various sugar particles (2011). Porous structure, in which the size and shape of the porogen particles determine the size and shape of the pores, as well as the morphology of the pores, is another important feature to consider when developing efficient oil absorbents [8]. Overall, sugar cubes are put in a container that will be utilized as a template for producing PDMS elastomers further on. In a 10:1 volume ratio, PDMS prepolymer and curing agent are prepared and applied on sugar cubes, which are subsequently degassed in a vacuum chamber to allow liquid prepolymer to penetrate between the sugar cubes. The PDMS-sugar cube is then cured for many hours. Finally, water extracts the sugar cubes from the PDMS template, resulting in a three-dimensional interconnected porous network of PDMS sponges.

A modified sugar leaching technique for PDMS sponge fabrication in *p*-xylene (PX) solution is shown in Figure 9.1 [18]. The PX solution aimed to make the PDMS prepolymer less viscous during the moulding process. As a result, enhancing its fluidity makes it possible to avoid bubbles and function without needing a vacuum. Carbon nanotubes have been introduced into the PDMS porous structure. The scarifying template used to improve surface roughness has been demonstrated by Ong and colleagues to produce high-surface-area polydimethylsiloxane (PDMS) that is extremely hydrophobic and oleophilic (2018) [19]. As an alternative to using sugar or salt as particulate leaching particles, Yu and colleagues developed a more direct and straightforward method for fabricating a PDMS sponge by using citric acid monohydrate (CAM) as a hard template and varying the mass ratio of 1:1, 1:3, and 1:7 to regulate the porosity structure of the sponge (2017). Since ethanol is super wetting and can penetrate the porous structure of PDMS, the CAM may be readily removed by dissolving it in

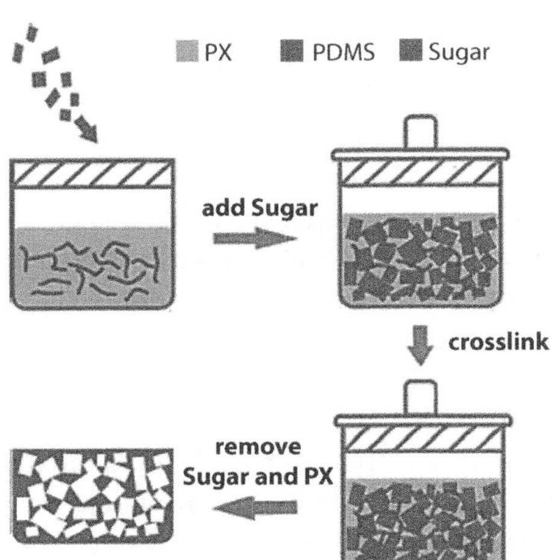

Figure 9.1 Schematic representation of the sugar template technique for producing PDMS oil absorbents [18] (adapted with permission).

CAM. When compared to polyethene foam or polypropylene foam, nickel foam offers a much more precise sacrificial template for the manufacture of PDMS sponges with well-defined size, shape, and wall width. As part of their research, Zhao and colleagues established a simplified method for producing oil-absorbing materials that comprised the use of candle shoot coating and PDMS modification, among other techniques. Consequently, a straightforward way for creating nickel foam surfaces with enhanced water and oil repulsion was established.

PDMS sponge materials are assessed for pore size and porosity using the sacrificial template approach. The size of the particles and the mixing ratio with a PDMS prepolymer solution are the two most prominent parameters to consider when employing the sacrificial template method. The pore size for sugar leaching is typically dozens to hundreds of micrometers. However, it can be reduced to tens of micrometers [17]. The use of additional sacrificial templates, such as colloidal crystal, as alternatives for sugar or salt, is necessary to compensate for the lack of narrow-sized pore distribution in these materials. When evaluating the porosity of PDMS sponges, the mixing ratio of particles and PDMS prepolymer solutions is crucial to the experiment's outcome. In association with a rise in the particle concentration in the prepolymer solution and a reduction in the density of the PDMS sponges, interconnected holes form [17].

9.2.2 Emulsion Templating Method

When the continuous emulsion phase is polymerized, emulsion templated macroporous polymers are formed, with the emulsion droplets acting as macropore templates [13]. Based on the internal phase volume ratio, which can be high (HIPE, >70 % or commonly 74 %), medium (MIPE, 30–70 %), or low (LIPE, 30 %), porous systems with varying pore size and porosity can be used. The products are referred to as polyHIPEs, polyMIPEs, and polyLIPEs, depending on their composition [17]. Because of its adaptability in changing material characteristics and the shape and thickness of the products, polymer emulsion templating is an intriguing approach for generating porous structures [20]. Controlling and understanding liquid emulsion template stability in ageing processes such as droplet coalescence is essential in establishing the optimum porous structure [17]. In order to determine the viability of using an emulsion templating strategy in the context of droplet microfluidics, it is necessary to consider several factors. To manufacture oil/water droplets, the surfaces must be sufficiently hydrophilic. Treatments must be long-lasting and irreversible, particularly when exposed to both oil and water [21].

This type of microporous polymer was generated in a continuous flow of water by combining water in oil emulsions, oil in water emulsions including ionic liquids, and supercritical carbon dioxide internal phases [13]. Water is one of the cheapest and most eco-friendly porogens among them [20]. Kovalenko *et al.*, for example, porous PDMS structure is created by polymerizing inverse water-in-oil emulsions (Using a crosslinking emulsion technique with repelling or adhesion interactions between water droplets, they developed bulk elastomer polyMIPE with isolated or linked porous structures [21]. Furthermore, because PDMS is a hydrophobic polymer, it has implications for droplet-based microfluidics. As a consequence, it is also ideal for generating droplets from a water-in-oil emulsion when the continuous phase needs to adequately moisten the device walls [21]. However, since the surfactant remains in the polymer, this technique has the disadvantage of not eliminating it. Therefore, surfactant elimination may be a time-consuming and difficult task. Emulsion templating, on the other hand, is frequently employed in 3D porous polymers for some applications such as tissue engineering, separation, water retention, and others [17]. Figure 9.2 shows the findings of research conducted by Riesco *et al.*, who demonstrated a three-step preparation technique for manufacturing centimeter-scale porous PDMS scaffolds with pores with diameters ranging from millimeter to micrometer in size could be constructed.

In the reticulation process, temperature and pressure have been used to regulate pore size and distribution in two distinct methods. In the first stage, deionized water (10% of the PDMS mass) was poured into the PDMS and agitated until the water-in-silicone emulsion reached 70%. Following that, the water-in-PDMS emulsion was used to form a cylindrical sample container, which was then filled with the sample. Vasquez *et al.*, created polyHIPE foams (PDMS foams) with open-cell and interconnected porous characteristics that are lightweight and durable using the HIPE templating

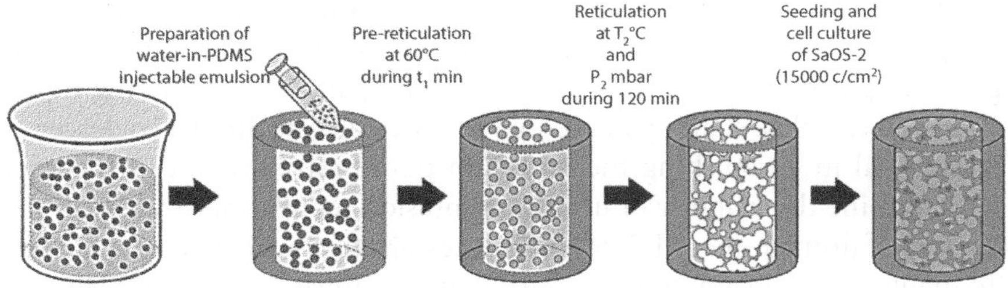

Figure 9.2 A schematic example of the PDMS porous scaffold production method employing water/PDMS emulsion casting [22] (adapted with permission).

technique (2020). The surface of the polyHIPE PDMS foams was coated with a layer of polydopamine (PDA). Then, they were dipped in a dopamine solution for 30 minutes. Ag nanoparticles were evenly dispersed using an *in-situ* growth method. PDMS foams were changed from hydrophobic to hydrophilic/underwater oleophilic due to the PDA layer's strong redox sites (hydroxyl and amine groups) and the chemical reduction of the adsorbed Ag precursors to generate Ag nanoparticles [23].

Based on a paper tissue, Sun *et al.*, recently produced superhydrophobic SiO_2/PDMS composites for oil absorption and water-in-oil emulsion separation to be used in oil filtration (2020). Paper tissue was chosen since it is a typical and inexpensive material with a natural capillary solid force that allows liquid traces to be quickly absorbed. PDMS/SiO_2 composites are produced using multiple processes as depicted in Figure 9.3a PDMS/SiO_2 paper coating dispersion is first generated, and then a paper tissue is submerged in the dispersion until the paper tissue is fully immersed. At the end of the process, the paper tissue is unfolded and allowed to dry at ambient temperature before being blasted at 120°C. for an hour. In a comparative investigation, ten different oils/organic solvents were studied, including motor oil, cyclohexane, butylene oxide, and others. As a consequence, the PDMS/SiO_2 composite was capable of absorbing roughly 20 times its weight in oils/organic solvents. Moreover, after 50 cycles, the composites paper's absorption capacity remained very steady. Due to the sheer paper tissue's strong capillary, PDMS/SiO_2 paper composites may absorb a broad spectrum of oils and organic solutions. These paper composites have the ability to independently extract oil from surfactant-stabilized water-in-oil emulsions while retaining high flux and separation efficiency throughout the extraction process [24].

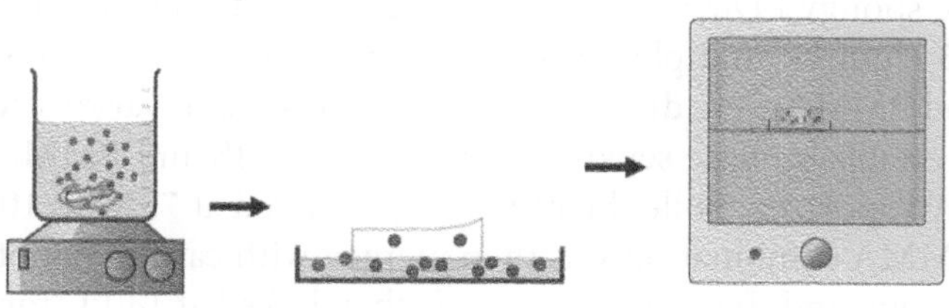

Figure 9.3 Three stages are involved in the fabrication of PDMS/SiO_2 composites paper: (a) Preparation of PDMS/SiO_2 paper coating dispersion. (b) Absorbing the dispersion with a paper tissue. (c) Unfolding the paper and allowing it to dry at room temperature before annealing it for about an hour at 120°C.

9.2.3 Phase Separation Method

The phase separation method is among the most promising technologies in the area of spongy polymer membrane manufacturing. It is one of the four major strategies, along with precipitation by cooling, which is also known as thermally induced phase separation (TIPS) [25]; nonsolvent-induced phase separation is the term used to describe the process of immersion precipitation (NIPS) [26]; vapor induced phase separation is defined as the precipitation of non-solvent (water) from the vapor phase by adsorption (VIPS) [27], and solvent evaporation-induced phase separation (EIPS) [28]. This approach is dependable and straightforward since the template dimensions do not restrict pore size. The templates do not need to be removed after the polymer has dried, making it one of the most convenient. Furthermore, because it is a surfactant-free process, the surfactant does not affect the mechanical properties of the polymer [29].

Among all of them, EIPS is the most commonly used technique for producing porous PDMS. This procedure entails dissolving a polymer in a volatile solvent and a less volatile non-solvent and then drying the resulting solution. As the solvent evaporates, the concentration of the polymer increases. When the solvent evaporated, the nonsolvent enriched droplets grew and merged, becoming more prominent. Separation of polymer-rich and polymer-lean phases is forced in the polymer solution [26]. The polymer-solvent system then transitions from the stable single-phase zone to the unstable two-phase zone through the metastable region, where nucleation and growth processes promote phase separation before returning to the stable single-phase zone. At the end of the process, removing the nonsolvent enriched droplets results in creating the porous structure [17].

To enhance PDMS porous morphology, Le *et al.* created porous structure PDMS-CNT nanocomposites by creating polymethylmethacrylate (PMMA) phase separation in PDMS using EIPS (2012). The manufacture of the spongy PDMS-CNT nanocomposite indicated schematically in Figure 9.3 requires multiple processes. To begin, different mixing ratios of PDMS and PMMA were dissolved in toluene at 75°C at various concentrations. To establish phase separation between PDMS and PMMA, toluene was slowly vaporized while the mixture was stirred at 75°C to 105°C. This PDMS/PMMA combination was then merged with carbon nanotubes and a curing agent, and the composition was then baked at 140°C for 12 hours to crosslink the PDMS and permanently fix the phase-separated structure. After being completely submerged in acetic acid to produce pores, the samples were thoroughly washed with deionized water to remove the PMMA present. When determining the pore size, factors, such as drying temperature and the quantities of three polymer components in the mixture, were critical to consider. A nitrogen gas blast and high-temperature drying were performed after the procedure to manufacture porous nanocomposites from the PDMS-CNT composite [14].

In addition, Abshirini *et al.* performed a further study to describe the fabrication and characterization of porous PDMS structures by utilizing the solvent EIPS (2021). To investigate the pore size distribution, mechanical properties, and porosity microstructure of the porous samples, a number of research techniques were used. In the PDMS-dense phase of the experiment, discrete holes with average pore diameters ranging from 330 to 1900 m were discovered. Raising the water/THF proportion while maintaining the PDMS percentage offers the greatest benefit of decreasing the average pore size of the water/THF combination. Nonetheless, raising the PDMS content from 30% to 60% while maintaining a steady water/THF ratio improved relative pore size [29].

9.2.4 3D Printing Techniques

Several improved techniques for 3D printing with PDMS have arisen due to advancements in additive manufacturing and the development of the maker movement [30]. Despite this, the liquid prepolymer's low elastic modulus enables direct 3D printing of PDMS in complicated geometries, which had previously been considered an unsolvable problem. There are two general methods to dealing with this problem. First, the starting material's thixotropic flow properties must be successfully fabricated by using low-temperature vulcanizing PDMS sealants, nonflowing PDMS elastomers, and PDMS prepolymers with increased viscosity, as well as fillers, to ensure that the material's geometric integrity is preserved during curing [17]. Due to the general low elastic modulus of pre-and post-cured PDMS, as well as its distortion under gravity, the 3D geometries that may be employed are highly restricted, resulting in structures that are only capable of supporting themselves in their whole [30].

Apart from that, conventional production processes (such as casting or weaving) have the following drawbacks moulds and every time the geometry of a product changes [31]. 3D printing has made it feasible to create large-area 3D structures without expensive moulds or multistep procedures in recent years due to significant advances in technology. A vast number of ordered and controlled structures may be readily created as a consequence of this process, and 3D printing technology, rather than the moulds often employed in prior fabrication techniques, allows for precise adjustment of the surface structure and wettability of the structures [32]. Additionally, unlike previous systems, this approach does not need the use of extra materials, such as sacrifice layers, or supplementary operations, such as the removal of the sacrificial layers and the insertion of the microneedles. Freeform deposition can be used for fabrication, and geometrical qualities can be easily adjusted by adjusting the printing parameters [33].

He *et al.* demonstrated a mould-free and straightforward strategy for manufacturing distinct wetting surfaces by designing an ordered porous structure with varying geometric properties using PDMS ink 3D printing in a single step (2017). The aligned filaments crossed perpendicularly over the underlying filaments, resulting in a typical anisotropic physical structure with multiple submillimeter grooves that proved ideal for generating anisotropic wettability using a computer program. Filament diameters (FD) drop as printing speeds rises from 0.75 to 6.00 mm/s. This is due to the nozzle's increased translation speed-boosting the stretching ratio of PDMS filaments. The superhydrophobic layers were also created by 3D printing PDMS ink into predetermined geometric designs, which eliminated the need for additional operations. Consequently, the enhanced wettability of the porous film created may be employed for air-breathable water-proofing applications in the future.

Duan and colleagues have also developed carbon nanomaterial-based SCMs that are extraordinarily stretchy and conductive. Split-level and aligned 3D porous polylactic acid (O-PLA) templates for O-PDMS were created using 3D printing. A 3D conducting network was established on the 3D frame when carbon nanofillers, such as carbon nanotubes (CNTs) and graphene were combined with the O-PDMS polymer. A-OPCG and S-OPCG, both OPDMS/CNTs/graphene composites with improved stretchability and conductivity, are the results of the last step in the process. The manufacturing procedure for S-OPCG is shown in Figure 9.4. Polylactic acid (PLA) was selected for printing the 3D skeleton since it is environmentally benign and easy to remove using a specific solvent.

Studies suggest that by altering the software and route of a 3D printer, they could produce structural O-PLA scaffolds with split-level and aligned geometries. After dissolving PLA and backfilling PDMS into the pores of O-PLA, the porous architectures of the material were precisely recreated, resulting in O-PDMS that was aligned and split-level in structure. As a consequence, in terms of electrical conductivity and deformation capability, S-OPCG surpassed A-OPCG. S-OPCG, for example, retains 40% of its conductivity after 100% uniaxial stretching, but A-OPCG only possesses 25%. Despite this, the electrical conductivity of S-OPCG remains constant after 5000 bending cycles and only marginally decreases after 100 repetitions of a % stretching-releasing process [34]. This method enables the

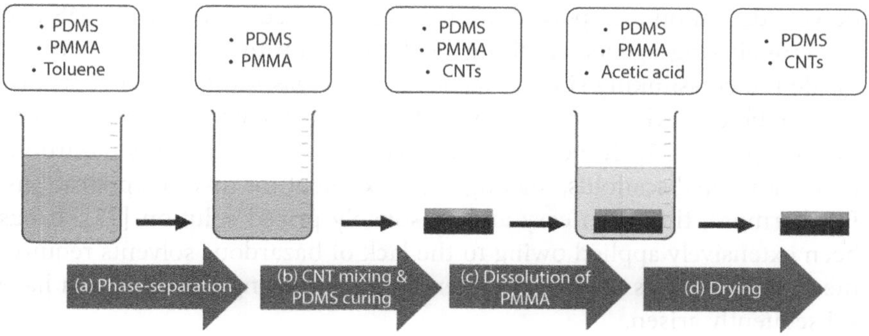

Figure 9.4 The schematic design for the manufacture of porous PDMS-CNT nanostructures.

creation of porous substrates with variable pore size, porosity, patterns, and a 3D gradient spongy look. However, this approach is time-consuming, costly, and pollutes the environment with hazardous organic solvents. As a result, a significant upscaling is no longer feasible. Additionally, this approach is still in its early stages, and further study is needed to improve its precision, economic feasibility, and dependability [17].

9.2.5 Gas-Forming Technique

The manufacture of polymer foams with open or closed pores for use in joint sealants, mechanical shock absorbers, insulators, and other applications is a well-known gas-forming technology. The process of forming gas is quite straightforward [17], and this method requires two simultaneous processes: linking reactions and foaming. The two most common silicone elastomer and rubber (SR) crosslinking procedures are vulcanization at high and low temperatures. High-temperature vulcanization SR systems were crosslinked using the radical reaction approach, and ethylene connections between the polysiloxane chains formed a polymer network [35]. To produce heat-activated vulcanization, a peroxide-vulcanizing agent is used to cure a higher molecular weight vinyl or methyl functionalized silicone polymer. This approach is known as "peroxide vulcanization" [36].

At room-temperature vulcanization systems, silanes, siloxanes, and polysiloxanes, which are crosslinking agents that attach to silicon atoms and comprise varied functional groups (hydroxyl, acetoxy, alkoxyl, and hydrogen atoms), as well as additional crosslinking agents, are required. In the presence of catalysts (mostly platinum or tin compounds), polycondensation and polyaddition reactions occur between PDMS and the crosslinking agents, resulting in the creation of a polymer network inside the crosslinking silicone substance. The gas is disseminated as discrete bubbles in closed-cell foam, and a polymer matrix generates a continuous phase [35]. In addition, gas foamed scaffolds are less hazardous than traditional scaffolds, making them excellent for in-vivo applications. Furthermore, this is an easy and reasonably priced solution [17]. It has been extensively applied owing to the lack of hazardous solvents requirement, medical uses of porous materials created using this approach have subsequently arisen.

According to Kobayashi *et al.*, a hydrosilylation cure for porous PDMS elastomers with porous structures formed by hydrogen foams may be achieved using hydrosilylation (1993). A vinyl group and a SiH group were used to make PDMS membranes, then treated with an -OH group to make them more flexible [37]. To produce porous biphasic calcium phosphate, Kim *et al.* used the foaming process (2012). In order to increase mechanical strength, they used hydroxyapatite (HA), and biodegradable BCP was made using tricalcium phosphate (TCP). The utilization of the foaming technique to create an exceptionally highly permeable wound dressing by polymerizing alcohol groups in the polyol and isocyanate groups in the toluene diisocyanate may have influenced the wound dressing's porosity, pore size, and mechanical properties. Toluene diisocyanate is used in the PU foaming method to form urethane bonds, when hydroxyl groups in the polyol react with the isocyanate groups in the toluene diisocyanate. In this method, the polyurea reaction, which occurs when the isocyanate groups react with H_2O as an additive, results in the formation of urea and CO_2 [38].

According to Li *et al.*, a viable technique for continuous PDMS/PVDF composite membrane fabrication was developed, as shown in Figure 9.5 (2019). As part of the coating slurry preparation process, the PDMS crosslinking agent and catalyst are mixed with an organic solvent to form a solution then sprayed onto the membrane support surface. A substantial volume of organic solvent must dissipate at room temperature to avoid explosion. After that the temperature is elevated to allow the PDMS covering layer to deep cure. The dried membrane is coiled up and placed on a take-up roll at the other end of the machine. When the linear PDMS is progressively crosslinked during the PDMS membrane manufacturing process, the coated layer moves from a liquid to a gel to a solid. As a consequence, to prevent sticking to surrounding membrane support, the PDMS coating layer must be dry before winding up. A small amount of catalyst was applied before coating in the presence of water to hinder fast cross-linking resulting in a lengthy curing process. As a result, the amount

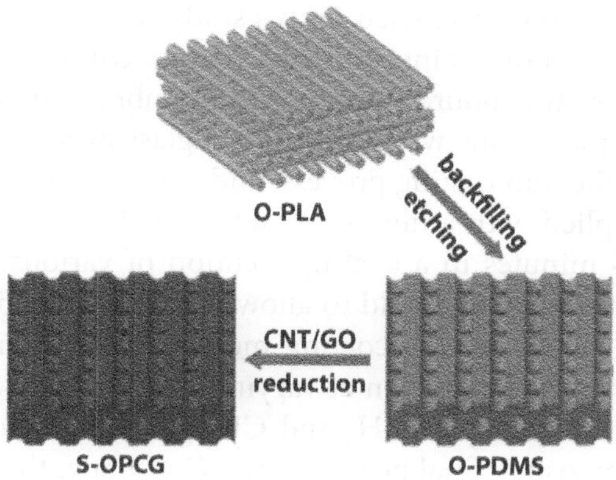

Figure 9.5 The schematic diagram of the preparation of S-OPCG [34] (adapted with permission).

of water used before the coating process was lowered. Ethanol was added to the PDMS coating slurry as an inhibitor to lower the temperature and slow down the crosslinking reaction. In order to facilitate the curing of the PDMS coating layer, higher temperatures were retained throughout the process. This was done by adding more catalysts to the coating solution [39]. An outline for the development of PDMS/PVDF-based composite membrane is illustrated in Figure 9.6.

Moreover, Kargari *et al.*, employed a foaming procedure to develop a polyetherimide (PEI) membrane covered with polydimethylsiloxane (PDMS) for hydrogen-methane separation (2014). PEI membranes were first dried at 105°C for 12 hours to eliminate any remaining solvents. To make uniform PDMS coating solutions, n-hexane was treated with PDMS polymer resin and crosslinking agent (in a weight ratio of 10:1 (w/w)).

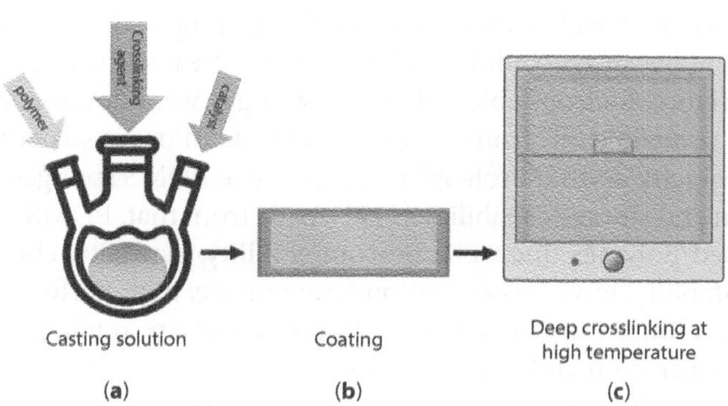

Figure 9.6 Manufacturing of PDMS/PVDF composite membrane [39].

Three coating methods were used in this study were: the 45-degree pouring of the solution, film casting, and dip-coating. Using the first approach, a coating solution was poured onto a PEI membrane that was angled at 45 degrees. The membrane was attached to a glass plate to prevent it from moving during the film casting process, and a 5-mm-thick layer of PDMS solution was applied with a film applicator. Finally, the membranes were immersed for 5 minutes in a coating solution of various concentrations before being withdrawn and held to allow the excess solution to drip out. This method is called the dip-coating method. After curing the created composite membrane in an oven at varying temperatures for 60 minutes, crosslinking would take place. H_2 and CH_4 gas permeation was used to evaluate the composites' final performance. Consequently, with escalating PDMS coating concentration, H_2 permeance initially decreased significantly but eventually stabilized [40].

9.3 PDMS Sponges as an Oil/Water Separation

PDMS sponge-based materials have been used in various sectors, such as flexible conductors [41], sensors [42], energy harvesting and storage devices [43], absorbents and oil/water separation appliances [4], biomedical applications [44], catalysis [45], microfluidics [46], and so on. The present chapter deals with the significant application of PDMS in oil/water separation.

Oil and other organic liquid pollutants endanger our ecology. When it comes to cleaning up oil spills in water, it includes three ways: collecting oil off the water's surface, combining oil with water, and burning it on the spot [4]. There are currently three primary methods up the pollution: collecting it off the ocean's surface, mixing it with water and utilizing dispersion agents to speed up natural degradation, and in situ burning [19]. By using oil-absorbent material to absorb spills, spill cleaning is simple, inexpensive, and efficient because no additional pollution is created during the process [18]. Therefore, hydrophobic polymeric sponges were developed recently to absorb reversible oils from oil spills over water. Porous structure's superhydrophobicity and superoleophilicity enhance PDMS sponges' oil/water separation and storage capabilities [17]. Aside from that, PDMS is the most widely used polymer due to its biocompatibility, elasticity, chemical and physical stability, low glass transition temperature, high performance, and ease of moulding [10]. As a result, its properties are retained even after repeated absorption and release cycles.

Zhang and his colleagues created a new oil-absorbent by combining the absorption processes of filling for porous materials and swell for organogels to boost their materials' absorption capacity for oil (2013). Interconnected pores and a stretchable structure allow these PDMS oil-absorbents to absorb and retain oils. To put it another way, the PDMS oil-absorbents were capable of absorbing up to 434 g/g of different oils and organic solvents. PDMS oil-absorbents are solid and flexible, and when squeezed, they recover more than 80% of the absorbed oil or organic solvent. Meanwhile, they can be used up to 20 times without losing their absorption capacity or weight [18]. That is the most important and attractive characteristic of PDMS.

Turco *et al.* mixed polydimethylsiloxane (PDMS) sponges with multi-walled carbon nanotubes (MWNTs) and a rigid template to generate a porous magnetic reusable nanocomposite with magnetic properties (2015). Magnetic sponges swell at a high and rapid rate of absorption in a variety of organic liquids. Organic liquids may also be quickly recovered by squeezing the sponges and re-used without deteriorating performance. To dilute the prepolymer, hexane was utilized as a solvent, and the absorption of dichloromethane was measured. According to the findings, PDMS-MWNT sponges with a volume ratio of 4:6 PDMS/hexane, a template particle size of 6829 m, and 3% (by weight) MWNTs on the template absorbed more than 70 g/g of dichloromethane [1].

The introduction of inorganic nanoparticles into a graphene/PDMS sponge resulted in more excellent mechanical stability, according to Pan *et al.* [47]. To lower the "trade-off" effect, a minimal quantity of graphene was deposited on the surface of a PDMS sponge skeleton, which improved mechanical durability and flexibility while simultaneously decreasing the concentration of graphene in the structure. Moreover, graphene on the surface of the PDMS sponge improves the quick adsorption of oils/organic solvents. This has resulted in a variation in maximum absorption capacity between 400% and 1500% of the weight of the PDMS sponge and graphene/PDMS sponge for various oils or organic solvents, depending on the material employed. Aside from that, after 100 cycles of compressive testing at an 80% strain under a 400-kPa load, graphene/PDMS can preserve its structure, suggesting that it has an outstanding mechanical elasticity and stability [47].

9.4 Conclusion

Porous PDMS materials with optimal pore shape and size have been developed to employ various methods and designs, including nature-inspired superhydrophobic materials. The convenience of manufacturing PDMS has been merged with the production process, including templates, substrates, and chemical treatments. A better understanding of existing manufacturing procedures for porous PDMS materials used in oil-water separation will assist in the creation of modular, high-performance materials that are based on novel engineering design and technical innovation. Adsorbents with rough surfaces and low-surface energy coating patterns are projected to perform better in oil-water separation, which will be beneficial to the environment as a result. The scalability of the manufacturing processes for porous PDMS sponges is crucial owing to the vast volume and big size of porous PDMS sponges required for large-scale production, especially for industrial applications. Low-cost, economically viable, and ecologically safe manufacturing processes should be used in their development.

References

1. Turco, A., Primiceri, E., Frigione, M., Maruccio, G., Malitesta, C., An innovative, fast and facile soft-template approach for the fabrication of porous PDMS for oil-water separation. *J. Mater. Chem. A*, 5, 45, 23785–23793, 2017.
2. Cojocaru, C., Macoveanu, M., Cretescu, I., Peat-based sorbents for the removal of oil spills from water surface: Application of artificial neural network modeling. *Colloids Surf. A Physicochem. Eng. Asp.*, 384, 1–3, 675–684, 2011.
3. Gupta, S. and Tai, N.H., Carbon materials as oil sorbents: A review on the synthesis and performance. *J. Mater. Chem. A*, 4, 5, 1550–1565, 2016.
4. Choi, S.J., Kwon, T.H., Im, H. *et al.*, A polydimethylsiloxane (PDMS) sponge for the selective absorption of oil from water. *ACS Appl. Mater. Interfaces*, 3, 12, 4552–4556, 2011.
5. Lu, Z., Song, J., Pan, K. *et al.*, EcoFlex sponge with ultrahigh oil absorption capacity. *ACS Appl. Mater. Interfaces*, 11, 22, 20037–20044, 2019.
6. Wu, J., Wang, N., Wang, L., Dong, H., Zhao, Y., Jiang, L., Electrospun porous structure fibrous film with high oil adsorption capacity. *ACS Appl. Mater. Interfaces*, 4, 6, 3207–3212, 2012.
7. Doshi, B., Sillanpää, M., Kalliola, S., A review of bio-based materials for oil spill treatment. *Water Res.*, 135, 262–277, 2018.
8. Halake, K., Bae, S., Lee, J. *et al.*, Strategies for fabrication of hydrophobic porous materials based on polydimethylsiloxane for oil-water separation. *Macromol. Res.*, 27, 2, 109–114, 2019.
9. González-Rivera, J., Iglio, R., Barillaro, G., Duce, C., Tinè, M.R., Structural and thermoanalytical characterization of 3D porous PDMS foam materials: The effect of impurities derived from a sugar templating process. *Polym. (Basel)*, 8, 6, 1–13, 2018.

10. Shin, J.H., Heo, J.H., Jeon, S., Park, J.H., Kim, S., Kang, H.W., Bio-inspired hollow PDMS sponge for enhanced oil–water separation. *J. Hazard. Mater.*, 365, 494–501, 2019.
11. Kim, D.H., Jung, M.C., Cho, S.H. *et al.*, UV-responsive nano-sponge for oil absorption and desorption. *Sci. Rep.*, 5, 1–12, 2015.
12. Zhao, F., Liu, L., Ma, F., Liu, L., Candle soot coated nickel foam for facile water and oil mixture separation. *RSC Adv.*, 4, 14, 7132–7135, 2014.
13. Tebboth, M., Jiang, Q., Kogelbauer, A., Bismarck, A., Inflatable elastomeric macroporous polymers synthesized from medium internal phase emulsion templates. *ACS Appl. Mater. Interfaces*, 7, 34, 19243–19250, 2015.
14. Lee, H., Yoo, J.K., Park, J.H., Kim, J.H., Kang, K., Jung, Y.S., A stretchable polymer-carbon nanotube composite electrode for flexible lithium-ion batteries: Porosity engineering by controlled phase separation. *Adv. Energy Mater.*, 2, 8, 976–982, 2012.
15. De Almeida Monteiro Melo Ferraz, M., Nagashima, J.B., Venzac, B., 3D printed mold leachates in PDMS microfluidic devices. *Sci. Rep.*, 10, 1–9, 2020.
16. Zargar, R., Nourmohammadi, J., Amoabediny, G., Preparation, characterization, and silanization of 3D microporous PDMS structure with properly sized pores for endothelial cell culture. *Biotechnol. Appl. Biochem.*, 63, 2, 190–199, 2016.
17. Zhu, D., Handschuh-Wang, S., Zhou, X., Recent progress in fabrication and aplication of polydimethylsiloxane sponges. *Microsyst. Nanoeng.*, 5, 16467–16497, 2017.
18. Zhang, A., Chen, M., Du, C., Guo, H., Bai, H., Li, L., Poly(dimethylsiloxane) oil absorbent with a three-dimensionally interconnected porous structure and swellable skeleton. *ACS Appl. Mater. Interfaces*, 5, 20, 10201–10206, 2013.
19. Ong, C.C., Murthe, S.S., Mohamed, N.M., Perumal, V., Saheed, M.S.M., Nanoscaled surface modification of poly(dimethylsiloxane) using carbon nanotubes for enhanced oil and organic solvent absorption. *ACS Omega*, 3, 11, 15907–15915, 2018.
20. Zhang, L., Zhang, Y., Chen, P., Du, W., Feng, X., Liu, B.F., Paraffin oil based soft-template approach to fabricate reusable porous PDMS sponge for effective oil/water separation. *Langmuir*, 35, 34, 11123–11131, 2019.
21. Trantidou, T., Elani, Y., Parsons, E., Ces, O., Hydrophilic surface modification of pdms for droplet microfluidics using a simple, quick, and robust method via pva deposition. *Microsyst. Nanoeng.*, 3, 1–9, April 2017.
22. Riesco, R., Boyer, L., Blosse, S. *et al.*, Water-in-PDMS emulsion templating of highly interconnected porous architectures for 3D cell culture. *ACS Appl. Mater. Interfaces*, 11, 32, 28631–28640, 2019.
23. Vásquez, L., Davis, A., Gatto, F. *et al.*, Multifunctional PDMS polyHIPE filters for oil-water separation and antibacterial activity. *Sep. Purif. Technol.*, 255, 117748, 2021.

24. Sun, J., Bi, H., Jia, H. *et al.*, A low cost paper tissue-based PDMS/SiO2 composite for both high efficient oil absorption and water-in-oil emulsion separation. *J. Clean. Prod.*, 244, 118814, 2020,
25. M'barki, O., Hanafia, A., Bouyer, D. *et al.*, Greener method to prepare porous polymer membranes by combining thermally induced phase separation and crosslinking of poly(vinyl alcohol) in water. *J. Membr. Sci.*, 458, 225–2235, 2014.
26. Chang, H.H., Beltsios, K., Lin, D.J., Cheng, L.P., Formation of polyamide 12 membranes via thermal-nonsolvent induced phase separation. *J. Appl. Polym. Sci.*, 130, 1, 14–24, 2013.
27. Tsai, H.A., Lin, J.H., Wang, D.M., Lee, K.R., Lai, J.Y., Effect of vapor-induced phase separation on the morphology and separation performance of polysulfone hollow fiber membranes. *Desalination*, 200, 1–3, 247–249, 2006.
28. Kim, J.K., Taki, K., Ohshima, M., Preparation of a unique microporous structure via two step phase separation in the course of drying a ternary polymer solution. *Langmuir*, 23, 24, 12397–12405, 2007.
29. Abshirini, M., Saha, M.C., Altan, M.C., Liu, Y., Cummings, L., Robison, T., Investigation of porous polydimethylsiloxane structures with tunable properties induced by the phase separation technique. *J. Appl. Polym. Sci.*, 138, 29, 1–13, 2021.
30. Hinton, T.J., Hudson, A., Pusch, K., Lee, A., Feinberg, A.W., 3D printing PDMS elastomer in a hydrophilic support bath via freeform reversible embedding. *ACS Biomater. Sci. Eng.*, 2, 10, 1781–1786, 2016.
31. Qin, Z., Compton, B.G., Lewis, J.A., Buehler, M.J., Structural optimization of 3D-printed synthetic spider webs for high strength. *Nat. Commun.*, 6, 1–7, May 2015.
32. He, Z., Chen, Y., Yang, J. *et al.*, Fabrication of polydimethylsiloxane films with special surface wettability by 3D printing. *Compos. B Eng.*, 129, 58–65, 2017.
33. Rahbar, M., Shannon, L., Gray, B.L. *et al.*, Corrigendum: Fabrication and characterization of a magnetic micro-actuator based on deformable Fe-doped PDMS artificial cilium using 3D printing. *Smart Mater. Struct.*, 24035015, 2015.
34. Duan, S., Yang, K., Wang, Z., Chen, M., Zhang, L., Zhang, H., Li, C., Fabrication of highly stretchable conductors based on 3D printed porous poly (dimethylsiloxane) and conductive carbon nanotubes/graphene network. *ACS Appl. Mater. Interfaces*, 8, 3, 2187–2192, 2016.
35. Yang, Z., Peng, H., Wang, W., Liu, T., Crystallization behavior of poly(ε-caprolactone)/layered double hydroxide nanocomposites. *J. Appl. Polym. Sci.*, 116, 5, 2658–2667, 2010.
36. Grande, J.B., Fawcett, A.S., McLaughlin, A.J., Gonzaga, F., Bender, T.P., Brook, M.A., Anhydrous formation of foamed silicone elastomers using the Piers-Rubinsztajn reaction. *Polym. (Guildf)*, 53, 15, 3135–3142, 2012.

37. Kobayashi, T., Saitoh, H., Fujii, N., Hoshino, Y., Takanashi, M., Porous membrane of polydimethylsiloxane by hydrosilylation cure: Characteristics of membranes having pores formed by hydrogen foams. *J. Appl. Polym. Sci.*, 50, 6, 971–979, 1993.
38. Kim, H.J., Park, I.K., Kim, J.H., Cho, C.S., Kim, M.S., Gas foaming fabrication of porous biphasic calcium phosphate for bone regeneration. *Tissue Eng. Regen. Med.*, 9, 2, 63–68, 2012.
39. Li, S., Li, P., Si, Z., Li, G., Qin, P., Tan, T., An efficient method allowing for continuous preparation of PDMS/PVDF composite membrane. *AIChE J.*, 65, 10, 1–13, 2019.
40. Kargari, A., Shamsabadi, A.A., Babaheidari, M.B., Influence of coating conditions on the H2 separation performance from H2/CH4 gas mixtures by the PDMS/PEI composite membrane. *Int. J. Hydrogen Energy*, 39, 12, 6588–6597, 2014.
41. Liang, S., Li, Y., Yang, J. *et al.*, 3D stretchable, compressible, and highly conductive metal-coated polydimethylsiloxane sponges. *Adv. Mater. Technol.*, 1, 7, 1–6, 2016.
42. Jung, S., Kim, J.H., Kim, J. *et al.*, Reverse-micelle-induced porous pressure-sensitive rubber for wearable human-machine interfaces. *Adv. Mater.*, 26, 28, 4825–4830, 2014.
43. McCall, W.R., Kim, K., Heath, C., La Pierre, G., Sirbuly, D.J., Piezoelectric nanoparticle polymer composite foams. *ACS Appl. Mater. Interfaces*, 6, 22, 19504–19509, 2014.
44. Pedraza, E., Brady, A.C., Fraker, C.A. *et al.*, Macroporous three-dimensional PDMS scaffolds for extrahepatic islet transplantation. *Cell Transplant.*, 22, 7, 1123–1125, 2013.
45. Li, X., Li, Y., Huang, Y. *et al.*, Organic sponge photocatalysis. *Green Chem.*, 19, 13, 2925–2930, 2017.
46. Zhou, T., Yang, J., Zhu, D. *et al.*, Hydrophilic sponges for leaf-inspired continuous pumping of liquids. *Adv. Sci.*, 4, 6, 1700028, 2017.
47. Pan, Z., Guan, Y., Liu, Y., Cheng, F., Facile fabrication of hydrophobic and underwater superoleophilic elastic and mechanical robust graphene/PDMS sponge for oil/water separation. *Sep. Purif. Technol.*, 261, 118273, 2021.

10

Polymer Nanocomposite-Based Anode for Bioelectrochemical Systems: A Review

Mohammad Danish Khan[1,2*], Abdul Hakeem Anwer[1] and Mohammad Zain Khan[1†]

[1]*Industrial Chemistry Research Laboratory, Department of Chemistry, Aligarh Muslim University, Aligarh, India*

[2]*Department of Chemical Engineering, Loughborough University, Loughborough, United Kingdom*

Abstract

An anode is an integral component of a bioelectrochemical system (BES), both electrochemically and practically. It supports bacterial biofilm formation and plays an essential role in the overall electrochemical reactions that are taking place in the reactor. Poor anode performance in BESs is another big issue for their reasonable applications. Therefore, modification of the anode is required to successfully improve the performance of BES. Recently, polymer nanocomposite-based anodes have been prepared and used in BES for the wastewater treatment and production of green bioenergy. Conducting polymer nanocomposites, such as polyaniline/graphene oxide and polypyrrole/MnO_2 were utilized as anode modifiers with different levels of success. These composites have a large specific electrode-surface area, economic accessibility, and an eco-friendly nature with superior performance similar to Pt electrodes. This chapter highlights a broad literature survey on the recent research progress made in BES anode modification utilizing polymer nanocomposites.

Keywords: Anode modification, bioelectrochemical system, green energy, polymer nanocomposite, wastewater treatment

**Corresponding author*: danishkhan2547upss@gmail.com

†*Corresponding author*: dr_mzain.fa@amu.ac.in

Shahid-ul-Islam, Abid Hussain Shalla and Mohammad Shahadat (eds.) Green Chemistry for Sustainable Water Purification, (223–242) © 2023 Scrivener Publishing LLC

10.1 Introduction

Nonrenewable energy is mostly derived from fossil fuels and is limited in its ability to meet the global increase in energy demand. Likewise, this source of energy is related to the emission of ozone-depleting substances (greenhouse gases) prompting natural contamination issues and global warming [1]. Fossil fuels, for example, natural gas and oil are estimated to be exhausted by the end of the year 2040 [2]. Thus, it is critically required to discover other renewable sources of energy. Choices are bioethanol, biodiesel, hydrogen fuel cell, biohydrogen [3–6], and bioenergy using bio-electrochemical system (BES) most common is microbial fuel cell (MFC) [1, 7] (Figure 10.1). The increasing expectations for everyday comforts and quickly developing population stimulate the worldwide demand for clean water, energy, and materials. The difficulties in the supply of energy and preservation, and in decreasing environmental contamination proficiency, make the advancement of BES technology, which can specifically use the energy of complex organic substrates with no other extra processing or strengthening the utilization of natural assets, vital in worldwide practical advancement [8]. BES as an encouraging alternative system for treating wastewater and self-managed energy generation using the metabolism of microbes have pulled strong attraction for scientists. This system has incredible possibilities for applications like bioelectricity generation [1, 7], treatment of wastewater [9, 10], bioremediation [1, 9, 10], biofuel production [6], and biosensors. A simple BES (MFC) setup includes ed anaerobic anodic chamber connected with a cathodic chamber (which is

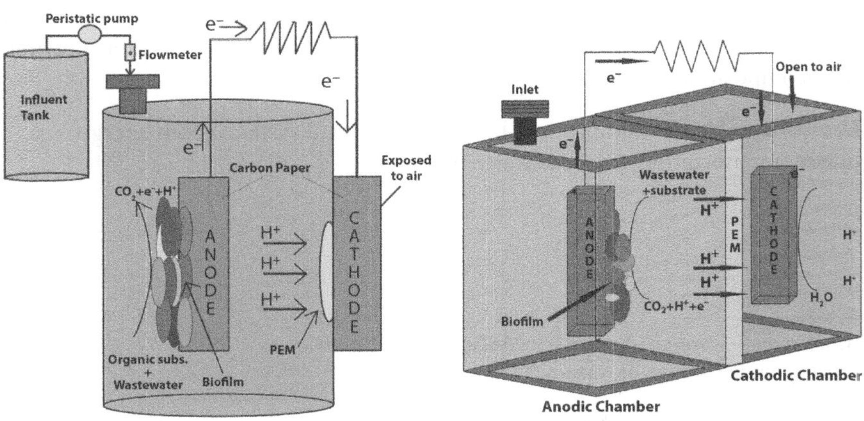

Figure 10.1 Single- and two-chamber BES.

open to air) by means of a proton or cation exchange membrane. The BES performance is directly influenced by many factors, such as ohmic loss, electron exchange from microbes using anode, microbial inoculum, type of substrate, electrode overpotential, and indirectly by a few factors like design of BES and cost of electrode and membrane [11]. Electrodes are an important part of BES whether it is in MFC for bioelectricity generation or microbial electrolysis cell (MEC) for generating biohydrogen and biofuel utilizing electricity [1, 6]. The anodic chamber is the basic segment for the BES system, as the microbial metabolic electrons are present in the anode chamber. As an essential part of BES, the material of anode has the highest effects on its execution. The morphology, architecture, and properties of the surface of anode materials have a significant impact on overall BES performance, such as affecting the attachment of microbes, oxidation of substrates, and electron transfer [11]. Even though BESs are environmentally friendly and have a built-in basic design, the qualities of the anode material play an important role and have been the root cause of low or even bad performance in some BES models. Ideal anode material must be favorable for microbial biofilm formation and also encourage the transfer and transport of electrons [12]. If microbial attachment on the anode is poor, it will suppress the productivity of transfer of an electron between bacteria and anode and ultimately it will affect the overall performance of BES and obstruct practical applications of BES technology [13]. Carbon felt (CF), carbon cloth (CC), carbon paper (CP), activated carbon fiber felt, graphite felt (GF), and graphite foil is the commonly used anodes in BES. Regardless of them being moderately cheap, stable, and showing great electronic conductivity, their innate character (hydrophobic) is harmful to vigorous microbes attachment bringing about low transfer of electron ability, and fouling of surface of materials from microbial discharge additionally intensified the issue [14, 15]. These are the factors responsible for the low energy yield of the unmodified anode. However, some properties (physicochemical) of the anodes could be improved utilizing artificially made materials using wanted characteristics to enhance the transfer of electron and microbial attachment to anode [16, 17]. Different modified anode materials have been functionalized in recent years with nanostructural modifications, for example, carbon nanotube (CNT), graphene and graphene oxide (GO), carbon black nanomaterial (CB), conducting polymers like polyaniline (PANI), polypyrrole (PPy), polymer nanocomposites like PANI/GO, PPy/GO, PANI/CNT [18]. To improve the efficiency of BES, valuable advancements have been made currently concerning anode modification. This chapter reviews and features the utilization of various anode materials for enhancing the overall BES performance and future directions in anode modification are talked about.

10.2 Conventional Anode Materials Based on Carbon

Due to properties, such as high surface area, higher conductivity for electricity, stability for chemicals, low cost, and biocompatibility, carbonaceous materials like CP, graphite brushes, CC, graphite rods, CF are the most commonly utilized anodes in BESs [19, 20]. In a work, Liu *et al.* [21] reported that by using a single compartment BES having 1 cathode and 8 graphite anodes, the achieved removal of almost 80% of chemical oxygen demand (COD), however, power density (PD) achieved was low due to low surface area and porosity of carbon-based anode. In other work, Chaudhuri *et al.* [22] used the same anode materials having a higher surface area (GF) and achieved three times greater current as reported by Liu and coworkers [21]. Logan and coworkers also used a graphite fiber brush as anode, and to increase surface area and microbial attachment to the anode, they coiled a titanium wire around the anode the maximum PD measured was 2400 mW/m^2, and it was four times greater than the PD measured using CP [23, 24]. CC also showed satisfactory PD. Due to the low cost of carbon mesh, Wang and coworkers used this as anode and PD achieved was higher than that of CC as anode [25, 26]. A new kind of filler anode was created through the mixing of CP and CF, which reduces the resistance by reducing grain limits and thus enhancing the association.

10.3 Modification of Anode with Nanomaterials Based on Carbon

Due to exceptional properties like specific surface area, high electronic conductivity, chemical stability, and strength, CNTs and graphene are famous carbon nanostructures. In small-sized BES, CNTs as modifying material for anode have shown great performance [27]. It is an effective material for anode modification in BES. Due to their outstanding stability for chemicals, biocompatibility, high conductivity, high specific area, and catalytic properties, CNTs have been used as promising substitute materials for BES anode [28]. CNTs also have strong growth properties, great cell attachment, and adhesion. In a work, Liang *et al.* [29] utilized CNT which is modified to improve biofilm attachment ultimately improving the performance of BES.

They showed that by using CNT powder as anode modifier with plain CP, the internal resistance of BES was less as compared to BES operated with only plain CP. Anode potential, generation of current, internal resistance in BES, and PD were affected by different loading rates of CNT powder. Thepsuparungsikul *et al.* [30] reported the influence of different morphologies of CNT like multiwalled CNT (MWCNT-COOH, MWCNT-OH) and single-walled (CNTCOOH) on the performance of BES performance. The BES operated with pore flow membrane MWCNT-OH has shown better performance as compared to other form of CNT and plain CC anode; this is because of the presence of OH functional group which improves the microbial attachment. Ren *et al.* [31] have utilized three types of CNT-modified anodes with various alignments and reported high PD in miniature BES. The BES operated with horizontally aligned CNT modified anode showed the lowest sheet resistance and thicker formation of biofilm as compared to BES worked with CNT which is randomly and vertically aligned [31]. Essentially, past research has demonstrated that CC anode altered with carbon nanomaterial doped with nitrogen expanded the anode ingestion of flavin discharged by *Shewanella oneidensis*, straightly promoting the electron transfer and improved contact area ensuing direct transfer of electrons from the microbes to the anode [32]. In a work, Erbay and coworkers reported the great charge transfer qualities shown by microbes grown on CNT due to a combination between pili and atoms of carbon [33]. To keep the ohmic resistances low in BES, CNTs can be applied directly over the stainless steel mesh. To establish a highly conductive anode for BES, CNTs can be applied over the CC to produce a large surface area and more PD than plain CC. Fraiwan and coworkers [34] demonstrated that CNT anodes will need modification of contact area for lessening the losses due to activation and poisoning of the cell. The advancement achieved in modification using chemical means and functional group attachment techniques for CNTs enhanced the action of surface, biocompatibility, accessibility, disclosing the path for an assortment of biological applications. A large portion of studies relates the enhancements of performance of BES with CNT modified two-dimensional anode based on carbon with the layer that is conductive that expanded the conductivity and contact area of the electrode. These examinations uncover that using two-dimensional CNTs modified anode can improve the capability of the microbes to bind and transfer electrons to the anode and reduce the internal resistance of BES [11]. Graphene is standout amongst the best nanomaterials with effective mechanical, electrical, and electricity conducting properties with high contact areas [11]. As of now graphene is rivaling CNT and in some cases surpass CNT in various implementations due to its low cost, high contact area, easy processing, and functionalization. GO, single and multilayer graphene, reduce graphene oxide (rGO) and hybridized nanocomposites are the family members of graphene nanomaterials [35]. Graphene is the most rigid and slimmest nanomaterial; its 2D grid made of hybridized carbon atoms and conjugation in long-range produces electrocatalytic properties [8]. This has initiated a solid logical and mechanical interest with extraordinary applications like energy storage devices, catalysts, nanoelectronics, biosensing. It might give new chances to BES also. Stainless steel mesh can also be modified using graphene [8]. In another work, Huang *et al.* [36] showed that by using a network of GO nanoribbons over an anode, the electron transfer process can be enhanced. These outcomes have been affirmed by Xiao *et al.* [37] where crumpled and regular graphene sheets and particles have been utilized for anode modification in BESs. The better execution of BES anode utilizing crusty particles of graphene has been credited to the larger surface area, higher electronic conductivity, and the free structure, which fundamentally upgrade mass transfer, biological reaction, and reaction kinetics. These outcomes have shown the strong capability of anode modified with graphene to enhance the execution of anode in BES. Additionally, by mixing graphene or CNTs with boron, sulfur, nitrogen, especially nitrogen functionals, electron transfer can be increased which ultimately improves the electrochemical behavior of BES [38]. Currently, the extraordinary qualities of graphene and CNTs have been used in the designing of three-dimensional anodes which are macroporous that empower inside accumulation and productive substrates transport, limiting the loss of energy in the BES [39]. In a work using a similar three-dimensional design, Flexer and coworkers [40] used reticulated vitreous carbon (RVC) and coated with carbon-based Nanoweb and accomplished a high current density (CD) compared with plain RVC anode, and the result was among the highest CD reported up to date carbon black nanomaterials (CBN) are broadly used to manufacture enzyme functionalized anodes as they possess characteristics appropriate to a bio-interface, i.e., relatively high surface, and contact area, high electrical conductivity combined with high porosity [41].

10.4 Metal or Metal Oxide-Based Modified Anode

Latest nanotechnology gives an exceptional chance to create a nanoparticle that alters anode as effective materials due to the extremely good structural, chemical, and electrical qualities of nanomaterials [11]. BES performance can be enhanced significantly by the alteration of the electrode with the help of a nanocomposite of metal or oxides of metals which reduces the hindrance of electron and enhance the electron transportation between the electrolyte & bacterial cell. A great number of researchers have explored the effect of variations in chemistry and morphology of the surface which introduced via nanomaterials, for example, metal and metal oxides on the overall execution of the BES [42, 43]. Different types of oxides of metals, such as titanium (Ti), iron (Fe), manganese (Mn), tin (Sn), etc. with carbonaceous material, have been used for the modification of anode [44]. A large number of metal oxides from which titanium (Ti) oxide has an alluring material that is structurally stable, electrically conductive, inexpensive & abundant in nature. A rare kind of physicochemical property shows a clear route for its implementation as an electrode modifier in BES. TiO_2 is also chemically stable & biologically compatible for the BES anode applications. Yin and coworkers [45] revealed that using TiO_2 nanotubes vertically adjusted on CP, the performance of BES improved. It was assigned to the development of 3D pores which is penetrating vertically offering a broad surface region for the coordinate exchange of electrons and more biocompatibility that encourages the diffusion of molecules and extremely good electron transfer channel. In a comparative report, Tang and coworkers [46] enhanced the anode execution of BES by utilizing egg white protein combined with TiO_2 and incorporated the mixture into a sponge framing a three-dimensional electrode with enhanced capacitance. TiO_2 independently did not drastically enhance the PD in BES because of the low electrical conductivity of TiO_2 operated with *Shewanella putrefaciens*. Therefore, it needs a modification along with CC or nickel. Performance of BES fused electrode with reduced graphene oxide (rGO) can be increased up to nineteen folds greater than the plain CC [47]. The glassy carbon electrode (GCE) was modified by MWCNT doped SnO_2. MWCNT-SnO_2/GCE in contrast to MWCNT/GCE composite alone or GCE alone in BES has a great impressive performance [48]. Hybrid electrode (SnO_2/rGO-CC) recovered almost five times more PD than its plain CC [49]. Zhu and coworkers [14] demonstrated that indium TiO_2 combined with nanosheets (GR) and poly (allylamine hydrochloride) improved the electrical conductive electron exchange between the bacteria and anode bringing about an increase in the level of bioelectricity. Oxides of iron, for example, magnetite (Fe_3O_4), hematite (Fe_2O_3), and goethite (Fe(OH)O) were utilized as modifiers of the electrode to enhance bacterial electron transfer with shifting level of achievement [16, 50, 51]. In a study, BES was modified by Fe_3O_4 composite with CNT in a mediator less BESs. This brings upon an attractive fascination that expanded the connection of anode with CNT, permitting the CNT to give a more ideal structural condition for the development of bacterial biofilm [14, 16]. By pyrolysis of ferrocene under normal pressure, Fe_2O_3 was coated on CC for reduction of cost, and this treatment was revealed that they increased the electron transfer efficiency of *S. oneidensis* based BES and also increased the maximum CD almost six times more than that of plain CC [50]. Peng *et al.* [51] showed a study where an enhancement in the capability of BES by 22% higher than that of normal electrode both kinetic activity and capacitance behavior of electrode by attachment of nano Fe_3O_4 to anodes. Similarly, the attachment of α-FeOOH (nanosemiconductor goethite) was studied by Peng and coworkers [51]. Here the PD has increased 36% as compared to that of the normal anode. This shows that the addition of nanosemiconductor goethite increases the extracellular transfer of electrons between electrodes and bacteria. The available surface region, electron exchangeability, and superb behavior of CP anode can be increased by the MnO_2 coating. MnO_2/MWCNT coating on a plain graphite electrode is utilized as an anode in BES [52]. The addition of 50% MnO_2 to MW-CNT enhanced the electrode's wettability and reduced the contact angle, resulting in improved kinetic activity and PD of BES operated with the modified electrode compared to its unmodified anode [52]. Alatraktchi and coworkers [53] showed the relationship between a microbial colony and nanomaterial density where the utilization of gold-based nanomaterial was demonstrated to upgrade BES performance by almost 30% compared to the unmodified counterpart.

10.5 Polymer-Based Modified Anode

Modification of anode can be done superbly by a polymer coating. A large number of studies showed the impact of conducting polymer modified anode on the performance of BES. Polyaniline (PANI) and polypyrrole (PPy) have got consideration in light of their great redox properties, high electrical conductivity, stability toward the environment, and simplicity of synthesis. In electrochemical studies, PANI has attracted more consideration due to its specific physical and chemical, capability of easily forming nanostructures, simplicity of preparation, and great stability [54, 55]. All the detailed outcomes showed that using PANI to modify the anode is a good technique to improve the performance of BESs. Different anodes modified with PANI have been targeted in BES, such as PANI-modified-CF, PANI-modified Pt-based CF, HSO_4^- combined PANI-modified CC, PANI-modified CF, and PANI modified glassy carbon [56]. By using PANI-modified natural loofah sponge nanostructured macroporous, the maximum PD achieved was almost 1100 mW/m^2 compared with graphene-coated sponge anode (612 mW/m^2) and RVC (650 mW/m^2) anodes. The enhancement of PD over those of ordinary three-dimensional anodes was an outcome of the synergetic impact of both parts: nitrogen-enhanced carbon nanoparticle covering that advanced extracellular electron transfer between anode and microbes and open three-dimensional structure that encouraged the development of biofilm. PPy due to its biocompatibility, stability, and high conductivity, has been utilized mostly in biomedical applications and biosensors even if the pH of the solution is neutral [57]. In a work, PPY-covered RVC was utilized as anode for BES prompting a huge improvement of energy generation in BES ascribed to the positive charge nature of PPy covering and the higher surface and contact area, which improved the electron transfer process [58]. In other work, granular PPY-fabricated anode and nanostructured fiber like PPY fabricated anode were utilized to enhance Bess's efficiency [59]. Results depicted an increment of 450% of PD using fiber like PPy-modified anode as compared to past reported outcomes.

10.6 Polymer Nanocomposites for Anode Modification

Certain drawbacks of PANI and PPy, for example, moderately low electrical conductivity, poor stability, and poor electron transfer properties restrict the BES performance if used in the long term. These drawbacks of conducting polymers expose the doors to utilizing the polymer nanocomposite modified anode as a powerful methodology for improving the performance of the anode in BES. Combining conductive polymers with nanomaterials (e.g., metal or metal oxides), the performance can be additionally improved which enhances the surface area for biofilm development and a number of catalytic sites. Additionally, CNT and graphene with polymer have been utilized to improve PD in BES due to their extraordinary properties. Incorporation of carbon nanostructures with conducting polymers is relied upon to enhance the performance of anode in BES due to the collective impact of both the materials, such as electrical conductivity, redox electrochemistry, and surface area. The use of nanocomposite based on PPy and PANI with CNTs as a modified BES anode resulted in improved PD in BES, which is a significant benefit for BES applications [60, 61]. Studies show that polymer nanocomposite-modified anode not only enhances the specific surface area and conductivity of anode but at the same time improves direct electron transfer, and consequently, exclude costly and other harmful mediators utilized in BES applications [62]. In graphene and conducting polymer-based nanocomposite, graphene serves as both doping and supporting agents. The examination of the monolithic and macroporous BES anode given PANI hybridized three-dimensional graphene demonstrated the fabricated anode outperforms the planar carbon anode as a result of its capabilities to 3D interface with bacterial biofilm, encourage electron transfer, and give exceptionally conductive pathways. Hou and colleagues utilized graphene/PANI nano complex-fabricated carbon cloth as a modified BES anode by combining the benefits of rGO and PANI. The electron transfer efficiency and loading of the bacterial biofilm were greatly improved by a graphene/PANI combination with a large surface area and conductivity. The enhancement has been credited to graphene because it provides a high surface area for PANI and also serves as a highly conductive support material [63]. CNTs are outstanding for their famous properties, for example, quality, hardness, remarkable conductivity, and lightweight. The PANI/CNT composites had a fiber-like design over the anodes and had a good discharge performance. The PANI/CNT which contains 20 wt% CNT displayed the maximum PD of 42 mW/m^2 and is credited to the predominant and specific catalytic effect of the PANI/CNT [59]. PPy/CNT composite exhibits high electrical conductivity and surface area which decreases the electron and mass transfer resistance and increases the contact between the electrode and the microorganisms. PPy/CNT brought about a PD of 228 mW/m^2 and good discharge performance [64]. For improving the performance of BES, PPy/rGO composites having high surface area and electrical conductivity have been used for CC modification [65]. Through electrostatic interaction, the positively charged PP_y improves the attachment of the negatively charged surface of the bacteria, encouraging the development of biofilm and better transfer of electrons. Using polymer nanocomposites-based anode, greater PD can be achieved as compared to unmodified PP_y. In another study, PPY/GO polymer nanocomposites on GF electrodes have been investigated as the anode of BES [66]. Compared to graphene or PPy alone, PPy/rGO with high surface area, high conductivity, stability, and biocompatibility improved the power production and BES performance in long term. Biocatalytic oxidation of glucose can be increased by providing more active sites, and high surface area [60].

Despite polymer nanocomposite with carbon-based nanostructures, research on conducting polymers demonstrate that combining metal or metal nanoparticles in a polymer framework produces nanocomposite materials of great sensing, optoelectronic, and catalysis properties [11]. In such a manner, PANI-TiO_2 nanocomposite has been used because of the cheap, nontoxic, stable, and biocompatible nature of TiO_2. Electrochemical degradation and cycle capability of PANI as electrode catalyst can be reduced by introducing TiO_2 in polymer matrix [67]. PANI/TiO_2 and Ti/TiO_2–PANI as anode for BES improved the power production and turned out to be reasonable materials for high power BES development [68]. Table 10.1 shows the various types of polymer nanocomposite materials used in BESs with corresponding efficiencies. In a work PANI–titania anodes were used in BESs with *E. coli* as a biocatalyst, which provides a large specific surface area for bacterial attachment due to the incorporation of titania nanostructures. *E. coli* cells adhered homogeneously to the surface of the anode with like hair structures. The pili allowed bacteria to adhere to the neighboring cells and showed a vital part as a mediator in the movement of the bacteria and biofilm formation, resulting in an improved PD of 1495 mW/m^2 [69]. These enhancements have been associated with high anode surface area, more biocompatibility, more electrochemical activity and

Table 10.1 Performance of BES (MFC) using modified polymer nanocomposite materials.

Type of BES	Carbon source	Modified anode	Efficiency	Reference
Two chamber	Acetate	PANI/MWCNT-GF	PD= 257 mW/m^2	[70]
Two chamber	Glucose	PANI/rGO/Pt-CC	PD= 2059 mW/m^2	[71]
Two chamber	Glucose	PPy/rGO-CC	PD= 1068 mW/m^2	[65]
Two chamber	Glucose	PANI/CNT-Ni foam	PD= 42 mW/m^2	[61]
Two chamber	Lactate	PPy/rGO-CF	PD= 1326 mW/m^2	[66]
Two chamber	Acetate	PANI/Ti-TiO_2-CF	PD= 980 mW/m^2	[68]
Two chamber	Acetate	PANI/GO-GF	PD= 183.8 mW/m^2	[72]
Two chamber	Azo dye wastewater	PPy/AQDS/Mn	PD= 84.58 mW/m^2	[73]

(*Continued*)

Table 10.1 Performance of BES (MFC) using modified polymer nanocomposite materials. (*Continued*)

Type of BES	Carbon source	Modified anode	Efficiency	Reference
Two chamber	Glucose	PANI/rGO-CC	PD= 1390 mW/m^2	[63]
Two chamber	Acetate	PANI/TiO_2-Ni foam	PD= 1490 mW/m^2	[69]
Two chamber	Glucose	PANI/Pt-GF	PD= 2900 mW/m^2	[74]
Two chamber	Glucose	PPy/CB-CP	PD= 452 mW/m^2	[58]
Two chamber	Acetate	PPy/MnO_2-GF	PD= 32.7 Wc/m^2	[75]
Two chamber	-	Lac. PANI/GO-GF	PD= 756 mW/m^2	[76]
Two chamber	Lactate	PANI/GO-GP	PD= 381 mW/m^2	[77]
Two chamber	Glucose	PPy/CNT-CP	PD= 650 mW/m^2	[78]
Two chamber	Lactate	PPy/GO-CF	CD= 18 A/m^2	[79]
Two chamber	Lactic zcid	PPy/AQDS-CF	PD= 1300 mW/m^2	[80]

stability, and giving a new condition for microbial attachment [70–80]. The incorporation of metal nanoparticles or carbon nanostructures in the polymers network is a successful method to improve the performance of anode and eliminating the drawbacks of conducting polymers [11].

It is clear from the literature that the growth of BESs with conducting polymer nanocomposites fabricated anodes is assumed to be higher as compared to other modifications in the coming future. Nanomaterials and polymer concentration play an essential role in the discharge cycles and cell potential determination and must be optimized before being used in BES [64].

10.7 Concluding Remarks and Future Perspectives

Currently, BESs are utilized in several applications, like bioelectricity generation, biofuel production, biosensors, bioremediation, and metal recovery. The impact of texture, composition, and properties of the surface of anode materials on the performance of BES must be examined clearly. Also, the working response of particular materials of anode for different microbes should be researched which will be useful for understudying the mechanism, depending on energy generation in BES. The standpoint for the use of BESs in sustainable energy generation might be very plausible because of the momentum level of the research exercise. The performance of BES has been incredibly upgraded in recent years by modifying the engineering and individual segments. The anode chamber is the most important part of BES since it exclusively decides the production and transportation of electrons. Anodic materials design and their techniques of modification are the most important deciding factors, which administer the overall performance of BESs. To deliver the greener BES, the materials developed from natural resources should be used as anodes. The current anode surface treatment methods are very costly and complex. Therefore, simple and strongly effective methods should be searched to advance these anodes. As of now, various techniques of surface modification are in progress like nanomaterials, treatment of the surface, polymer, polymer nanocomposites to understand the performance and compatibility issue. Studies concentrated on the modifications serve to the other options to the bioincompatible metal anodes that would positively improve the BES energy generation. Nanoparticle's arrangement is of the field of interest in recent years. The application of nanomaterials over anode by utilizing electrochemistry is likewise dynamic. Therefore, more efforts should be given to unite forward those innovations. The impact of morphology, structure, and aspect of conducting polymer nanocomposite catalysts over the performance of BES must be studied in detail. Besides the nanometal catalysts, bi, tri, and quadratic metal catalysts should be planned and arranged to improve the surface and contact area and number of active sites of polymer nanocomposites to enhance the catalytic fuel oxidation. Positively charged functional groups are introduced on the anode surface after modification which ultimately interacts with a negatively charged cell of bacteria. Unfortunately, across the broad use, BES is yet to be practically inferable from the cost of material and further intensified by the way that modification generally causes environmental issues. Additionally, research and analysis are needed to be explored to set up functional group stability on the anode surface that can improve the generation of power and long-term operation of BES. The demanding R&D of nanotechnology & material science is required to develop novel polymer nanocomposites with exceptional mechanical, physical, and chemical properties, which will bring critical advancement to energy-based applications. Despite significant achievements in the architecture of nanocomposites, there are still many challenges and problems that need to be addressed for their application in BES. For their application in BES, numerous factors must be considered like material choice and accessibility, conductivity, synthesis methods, biocompatibility, and surface area. Advancements in material science and nanotechnology may give novel devices to the improvement of BES anode applications in the future.

References

1. Khan, M.D., Khan, N., Sultana, S. *et al.*, Bioelectrochemical conversion of waste to energy using microbial fuel cell technology. *Process Biochem.*, 57, 141–58, 2017.
2. Shafiee, S. and Topal, E., When will fossil fuel reserves be diminished? *Energy Policy*, 37, 181–9, 2009.
3. Ahmed, E.M., Hydrogel: Preparation, characterization, and applications: A review. *J. Adv. Res.*, 6, 105–21, 2015.
4. Parmar, A., Singh, N.K., Pandey, A., Gnansounou, E., Madamwar, D., Cyanobacteria and microalgae: A positive prospect for biofuels. *Bioresour. Technol.*, 102, 10163–72, 2011.
5. Sin, M.C., Gan, S.N., Annuar, M.S.M., Tan, I.K.P., Thermodegradation of medium-chain-length poly(3-hydroxy alkanoates) produced by pseudomonas putida from oleic acid. *Polym. Degrad. Stab.*, 95, 2334–42, 2010.
6. Khan, M.Z., Nizami, A.S., Rehan, M., Ouda, O.K.M., Sultana, S., Ismail, I.M *et al.*, Microbial electrolysis cells for hydrogen production and urban wastewater treatment: A case study of Saudi Arabia. *Appl. Energy*, 185, 410–20, 2017.
7. Khan, M.D., Abdulateif, H., Ismail, I.M., Sabir, S., Khan, M.Z., Logan, B. *et al.*, Bioelectricity generation and bioremediation of an azo-dye in a microbial fuel cell coupled activated sludge process. *PloS One*, 10, e0138448, 2015.
8. Zhang, Y., Liu, L., Van der Bruggen, B., Yang, F., Nanocarbon based composite electrodes and their application in microbial fuel cells. *J. Mater. Chem. A*, 5, 12673–98, 2017.
9. Sultana, S., Khan, M.D., Sabir, S., Gani, K.M., Oves, M., Khan, M.Z., Bioelectro degradation of azo-dye in a combined anaerobic–aerobic process along with energy recovery. *New J. Chem.*, 39, 9461–70, 2015.

10. Khan, M.Z., Singh, S., Sultana, S., Sreekrishnan, T.R., Ahammad, S.Z., Studies on the biodegradation of two different azo dyes in bioelectrochemical systems. *New J. Chem.*, 39, 5597–604, 2015.
11. Dumitru, A. and Scott, K., *Anode materials for microbial fuel cells*, Elsevier Ltd, United Kingdom, 2015.
12. Ci, S., Wen, Z., Chen, J., He, Z., Decorating anode with bamboo-like nitrogen-doped carbon nanotubes for microbial fuel cells. *Electrochem. Commun.*, 14, 71–4, 2012.
13. Song, Y.C., Kim, D.S., Woo, J.H., Subha, B., Jang, S.H., Sivakumar, S., Effect of surface modification of anode with surfactant on the performance of microbial fuel cell. *Int. J. Energy Res.*, 39, 860–8, 2015.
14. Zhu, Y., Ji, J., Ren, J., Yao, C., Ge, L., Conductive multilayered polyelectrolyte films improved performance in microbial fuel cells (MFCs). *Colloids Surfaces A. Physicochem. Eng. Asp.*, 455, 92–6, 2014.
15. Chou, H.T., Lee, H.J., Lee, C.Y., Tai, N.H., Chang, H.Y., Highly durable anodes of microbial fuel cells using a reduced graphene oxide/carbon nanotube-coated scaffold. *Bioresour. Technol.*, 169, 532–6, 2014.
16. Park, I.H., Christy, M., Kim, P., Nahm, K.S., Enhanced electrical contact of microbes using Fe_3O_4/CNT nanocomposite anode in mediator-less microbial fuel cell. *Biosens. Bioelectron.*, 58, 75–80, 2014.
17. Kang, C.S., Eaktasang, N., Kwon, D.Y., Kim, H.S., Enhanced current production by desulfovibrio desulfuricans biofilm in a mediator-less microbial fuel cell. *Bioresour. Technol.*, 165, 27–30, 2014.
18. Nosek, D., Jachimowicz, P., Kwiatkowska, A.C., Anode modification as an alternative approach to improve electricity generation in microbial fuel cells. *Energies*, 13, 6596, 2020.
19. Khan, M.D., Li, D., Tabraiz, S. *et al.*, Integrated iron phthalocyanine supported air cathode microbial fuel cell-aerobic bioreactor set-up for enhanced bioelectrodegradation of azo dye acid blue 29. *Sci. Total Environ.*, 756, 143752, 2021.
20. Khan, N., Anwer, A.H., Khan, M.D., Ibhadon, A., Azam, A., Khan, M.Z., Magnesium ferrite spinels as anode modifier for the treatment of Congo red and energy recovery in a single chambered microbial fuel cell. *J. Hazard. Mater.*, 124561, 2020.
21. Liu, H., Cheng, S., Logan, B.E., Power generation in fed-batch microbial fuel cells as a function of ionic strength, temperature, and reactor configuration. *Environ. Sci. Technol.*, 39, 5488–93, 2005.
22. Chaudhuri, S.K. and Lovley, D.R., Electricity generation by direct oxidation of glucose in mediatorless microbial fuel cells. *Nat. Biotechnol.*, 21, 1229–32, 2003.
23. Ahn, Y. and Logan, B.E., Effectiveness of domestic wastewater treatment using microbial fuel cells at ambient and mesophilic temperatures. *Bioresour. Technol.*, 101, 469–75, 2010.

24. Logan, B., C.heng, S., Watson, V., Estadt, G., Graphite fiber brush anodes for increased power production in air-cathode microbial fuel cells. *Environ. Sci. Technol.*, 41, 3341–6, 2007.
25. Wang, X., Feng, Y.J., Lee, H., Electricity production from beer brewery wastewater using single chamber microbial fuel cell. *Water Sci. Technol.*, 57, 1117–21, 2008.
26. Wang, X., Cheng, S., Feng, Y., Merrill, M.D., Saito, T., Logan, B.E., Use of carbon mesh anodes and the effect of different pretreatment methods on power production in microbial fuel cells. *Environ. Sci. Technol.*, 43, 6870–4, 2009.
27. Kou, T., Yang, Y., Yao, B., Li., Y., Interpenetrated bacteria-c nanotubes film for microbial fuel cells. *Small Methods*, 1800152, 1–6, 2018.
28. Regan, B., Gratzel, M., low-cost, A., High-efficiency solar cell based on dye-sensitized colloidal TiO2 films. Nature publishing group. *Nature*, 353, 737–40, 1991.
29. Liang, P., Wang, H., Xia, X., Huang, X., Mo, Y., Cao, X et al, Carbon nanotube powders as electrode modifier to enhance the activity of anodic biofilm in microbial fuel cells. *Biosens. Bioelectron.*, 26, 3000–4, 2011.
30. Thepsuparungsikul, N., Ng, T.C., Lefebvre, O., Ng, H.Y., Different types of carbon nanotube-based anodes to improve microbial fuel cell performance. *Water Sci. Technol.*, 69, 1900–10, 2014.
31. Ren, H., Pyo, S., Lee, J.I., Park, T.J., Gittleson, F.S., Leung, F.C.C. *et al.*, A high power density miniaturized microbial fuel cell having carbon nanotube anodes. *J. Power Sources*, 273, 823–30, 2015.
32. Yu, Y.Y., Guo, C.X., Yong, Y.C., Li, C.M., Song, H., Nitrogen doped carbon nanoparticles enhanced extracellular electron transfer for high-performance microbial fuel cells anode. *Chemosphere*, 140, 26–33, 2015.
33. Erbay, C., Pu, X., Choi, W., Choi, M., Ryu, Y., Hou, H. *et al.*, Control of geometrical properties of carbon nanotube electrodes toward high-performance microbial fuel cells. *J. Power Sources*, 280, 347–54, 2015.
34. Fraiwan, A., Adusumilli, S.P., Han, D., Steckl, A.J., Call, D.F., Westgate, C.R. *et al.*, Microbial power-generating capabilitie on micro-/nano-structured anodes in micro-sized microbial fuel cells. *Fuel Cells*, 14, 801–9, 2014.
35. Dideikin, A.T. and Vul, A.Y., Graphene oxide and derivatives: The place in graphene family. *Front. Phys.*, 6, 149, 2019.
36. Huang, Y.X., Liu., X.W., Xie, J.F., Sheng, G.P., Wang, G.Y., Zhang, Y.Y. *et al.*, Graphene oxide nanoribbons greatly enhance extracellular electron transfer in bio-electrochemical systems. *Chem. Commun.*, 47, 5795, 2011.
37. Xiao, L., Damien, J., Luo, J., Jang, H.D., Huang, J., He, Z., Crumpled graphene particles for microbial fuel cell electrodes. *J. Power Sources*, 208, 187–92, 2012.
38. Xu, Z., Li, H., Fu, M., Luo, H., Sun, H., Zhang, L. *et al.*, Nitrogen-doped carbon nanotubes synthesized by pyrolysis of nitrogen-rich metal phthalocyanine derivatives for oxygen reduction. *J. Mater. Chem.*, 22, 18230, 2012.

39. Rahman, G. and Najaf, Z., An overview of the recent progress in the synthesis and applications of carbon nanotubes. *J. Carbon Res.*, 5, 3, 2019.
40. Flexer, V., Chen, J., Donose, B.C., Sherrell, P., Wallace, G.G., Keller, J., The nanostructure of three-dimensional scaffolds enhances the current density of microbial bioelectrochemical systems. *Energy Environ. Sci.*, 6, 1291, 2013.
41. Minteer, S.D., Atanassov, P., Luckarift, H.R., Johnson, G.R., New materials for biological fuel cells. *Mater Today*, 15, 166–73, 2012.
42. Lv, Z., Xie, D., Yue, X., Feng, C., Wei, C., Ruthenium oxide-coated carbon felt electrode: A highly active anode for microbial fuel cell applications. *J. Power Sources*, 210, 26–31, 2012.
43. Fan, Y., Xu, S., Schaller, R., Jiao, J., Chaplen, F., Liu, H., Nanoparticle decorated anodes for enhanced current generation in microbial electrochemical cells. *Biosens. Bioelectron.*, 26, 1908–12, 2011.
44. Hindatu, Y., Annuar, M.S.M., Gumel, A.M., Mini-review: Anode modification for improved performance of microbial fuel cell. *Renew. Sustain. Energy Rev.*, 73, 236–48, 2017.
45. Yin, T., Lin, Z., Su, L., Yuan, C., Fu, D., Preparation of vertically oriented TiO_2 nanosheets modified carbon paper electrode and its enhancement to the performance of MFCs. *ACS Appl. Mater. Interfaces*, 7, 400–8, 2015.
46. Tang, J., Yuan, Y., Liu, T., Zhou, S., High-capacity carbon-coated titanium dioxide core-shell nanoparticles modified three dimensional anodes for improved energy output in microbial fuel cells. *J. Power Sources*, 274, 170–6, 2015.
47. Zou, L., Qiao, Y., Wu, X.S., Ma, C.X., Li, X., Li, C.M., Synergistic effect of titanium dioxide nanocrystal/reduced graphene oxide hybrid on enhancement of microbial electrocatalysis. *J. Power Sources*, 276, 208–14, 2015.
48. Mehdinia, A., Ziaei, E., Jabbari, A., Multi-walled carbon nanotube/SnO_2 nanocomposite: A novel anode material for microbial fuel cells. *Electrochim. Acta*, 130, 512–8, 2014.
49. Mehdinia, A., Ziaei, E., Jabbari, A., Facile microwave-assisted synthesized reduced graphene oxide/tin oxide nanocomposite and using as anode material of microbial fuel cell to improve power generation. *Int. J. Hydrog. Energy*, 39, 10724–30, 2014.
50. Zhou, S., Tang, J., Yuan, Y., Conduction-band edge dependence of carbon-coated hematite stimulated extracellular electron transfer of shewanella oneidensis in bioelectrochemical systems. *Bioelectrochemistry*, 102, 29–34, 2015.
51. Peng, X., Yu, H., Wang, X., Gao, N., Geng, L., Ai, L., Enhanced anode performance of microbial fuel cells by adding nanosemiconductor goethite. *J. Power Sources*, 223, 94–9, 2013.
52. Fu, Y., Yu, J., Zhang, Y., Meng, Y., Graphite coated with manganese oxide/multiwall carbon nanotubes composites as anodes in marine benthic microbial fuel cells. *Appl. Surf. Sci.*, 317, 84–9, 2014.

53. Alatraktchi, F.A.Z., Zhang, Y., Angelidaki, I., Nanomodification of the electrodes in microbial fuel cell: Impact of nanoparticle density on electricity production and microbial community. *Appl. Energy*, 116, 216–22, 2014.
54. Qin, Y., Alginate fibers: An overwiew of the production processes and applications in wound management. *Polym. Int.*, 57, 171–80, 2008.
55. Huang, Z., Liu, E., Shen, H., Xiang, X., Tian, Y., Xiao, C. *et al.*, Preparation of polyaniline nanotubes by a template-free self-assembly method. *Mater. Lett.*, 65, 2015–8, 2011.
56. Lai, B., Tang, X., Li, H., Du, Z., Liu, X., Zhang, Q., Power production enhancement with a polyaniline modified anode in microbial fuel cells. *Biosens. Bioelectron.*, 28, 373–7, 2011.
57. Balint, R., Cassidy, N.J., Cartmell, S.H., Conductive polymers: Towards a smart biomaterial for tissue engineering. *Acta Biomater.*, 10, 2341–53, 2014.
58. Yuan, Y. and Kim, S., Polypyrrole-coated reticulated vitreous carbon as anode in microbial fuel cell for higher energy output. *Fuel Cell*, 29, 168–72, 2008.
59. Zou, Y., Pisciotta, J., Baskakov, I.V., Nanostructured polypyrrole-coated anode for sun-powered microbial fuel cells. *Bioelectrochemistry*, 79, 50–6, 2010.
60. Zou, Y., Xiang, C., Yang, L., Sun, L.X., Xu, F., Cao, Z., A mediatorless microbial fuel cell using polypyrrole coated carbon nanotubes composite as anode material. *Int. J. Hydrog. Energy*, 33, 4856–62, 2008.
61. Qiao, Y., Li, C.M., Bao, S.J., Bao, Q.L., Carbon nanotube/polyaniline composite as anode material for microbial fuel cells. *J. Power Sources*, 170, 79–84, 2007.
62. Upadhyayula, V.K.K. and Gadhamshetty, V., Appreciating the role of carbon nanotube composites in preventing biofouling and promoting biofilms on material surfaces in environmental engineering: A review. *Biotechnol. Adv.*, 28, 802–16, 2010.
63. Hou, J., Liu, Z., Zhang, P., A new method for fabrication of graphene/polyaniline nanocomplex modified microbial fuel cell anodes. *J. Power Sources*, 224, 139–44, 2013.
64. Kumar, G.G., Sarathi, V.G.S., Nahm, K.S., Recent advances and challenges in the anode architecture and their modifications for the applications of microbial fuel cells. *Biosens. Bioelectron.*, 43, 461–75, 2013.
65. Kumar, G.G., Kirubaharan, C.J., Udhayakumar, S., Ramachandran, K., Karthikeyan, C., Renganathan, R. *et al.*, Synthesis, structural, and morphological characterizations of reduced graphene oxide-supported polypyrrole anode catalysts for improved microbial fuel cell performances. *ACS Sustain. Chem. Eng.*, 2, 2283–90, 2014.
66. Lv, Z., Chen, Y., Wei, H., Li, F., Hu, Y., Wei, C. *et al.*, One-step electrosynthesis of polypyrrole/graphene oxide composites for microbial fuel cell application. *Electrochim. Acta*, 111, 366–73, 2013.

67. Bian, C., Yu, A., Wu, H., Fibriform polyaniline/nano-TiO_2 composite as an electrode material for aqueous redox supercapacitors. *Electrochem. Commun.*, 11, 266–9, 2009.
68. Benetton, X.D., Navarro-Avila, S., Carrera-Figueiras, C., Electrochemical evaluation of Ti/TiO_2-polyaniline anodes for microbial fuel cells using hypersaline microbial consortia for synthetic-wastewater treatment. *J. New Mater Electrochem. Syst.*, 13, 1–6, 2010.
69. Qiao, Y., Bao, S., Li, C.M., Cui, X., Lu, Z., Guo, J., Nanostructured polyaniline/titanium fuel cells. *ACS Nano*, 2, 113–9, 2008.
70. Cui, H.F., Du, L., Guo, P.B., Zhu, B., Luong, J.H.T., Controlled modification of carbon nanotubes and polyaniline on macroporous graphite felt for high-performance microbial fuel cell anode. *J. Power Sources*, 283, 46–53, 2015.
71. Kumar, G.G., Kirubaharan, C.J., Udhayakumar, S., Karthikeyan, C., Nahm, K.S., Conductive polymer/graphene supported platinum nanoparticles as anode catalysts for the extended power generation of microbial fuel cells. *Ind. Eng. Chem. Res.*, 53, 16883–93, 2014.
72. Jiang, X., Lou, S., Chen, D., Shen, J., Han, W., Sun, X. *et al.*, Fabrication of polyaniline/graphene oxide composite for graphite felt electrode modification and its performance in the bioelectrochemical system. *J. Electroanal. Chem.*, 744, 95–100, 2015.
73. Sun, J., Cai, B., Xu, W., Huang, Y., Zhang, Y., Peng, Y. *et al.*, Enhanced bioelectricity generation and azo dye treatment in a reversible photo-bioelectrochemical cell by using novel anthraquinone-2,6-disulfonate (AQDS)/MnOx-doped polypyrrole film electrodes. *Bioresour. Technol.*, 225, 40–7, 2017.
74. Prasad, D., Arun, S., Murugesan, M., Padmanaban, S., Satyanarayanan, R.S., Berchmans, S. *et al.*, Direct electron transfer with yeast cells and construction of a mediatorless microbial fuel cell. *Biosens. Bioelectron.*, 22, 2604–10, 2007.
75. Chen, D., Shen, J., Jiang, X., Mu, Y., Ma, D., Han, W. *et al.*, Fabrication of polypyrrole/β-MnO_2 modified graphite felt anode for enhancing recalcitrant phenol degradation in a bioelectrochemical system. *Electrochim. Acta*, 244, 119–28, 2017.
76. Hui, J., Jiang, X., Xie, H., Chen, D., Shen, J., Sun, X. *et al.*, Laccase-catalyzed electrochemical fabrication of polyaniline/graphene oxide composite onto graphite felt electrode and its application in bioelectrochemical system. *Electrochim. Acta*, 190, 16–24, 2016.
77. Sun, D.Z., Yu, Y.Y., Xie, R.R., Zhang, C.L., Yang, Y., Zhai, D.D. *et al.*, *In-situ* growth of graphene/polyaniline for synergistic improvement of extracellular electron transfer in bioelectrochemical systems. *Biosens. Bioelectron.*, 87, 195–202, 2017.

78. Khan, N., Anwer, A.H., Ahmad, A. *et al.*, Investigation of CNT/PPy-modified carbon paper electrodes under anaerobic and aerobic conditions for phenol bioremediation in microbial fuel cells. *ACS Omega*, 5, 471–480, 2020.
79. Wang, H., Liu, Y., Li, M., Huang, H., Xu, H.M., Hong, R.J. *et al.*, Multifunctional TiO_2 nanowires-modified nanoparticles bilayer film for 3D dye-sensitized solar cells. *Optoelectron. Adv. Mater. Rapid Commun.*, 4, 1166–9, 2010.
80. Feng, C., Ma, L., Li, F., Mai, H., Lang, X., Fan, S., A polypyrrole/anthraquinone-2,6-disulphonic disodium salt (PPy/AQDS)-modified anode to improve performance of microbial fuel cells. *Biosens. Bioelectron.*, 25, 1516–20, 2010.

11

Electrospinning Setup Design and Modification for Fabrication of Photocatalytic Electrospun Nanofibrous Membranes for Water Treatment

N. Awang[1]*, A.M. Nasir[2,3], S.J. Fatihhi[1], A. Johari[1], S. Shaharuddin[1], A.H. Bakri[1], M.F.M. Alkbir[1], M.A.M. Yajid[2,4] and J. Jaafar[2,3]

[1]Plant Engineering Technology (PETech), Malaysian Institute of Industrial Technology, Universiti Kuala Lumpur, Persiaran Sinaran Ilmu, Bandar Seri Alam, Johor, Malaysia
[2]Faculty of Engineering, Universiti Teknologi Malaysia, UTM Skudai, Johor Bahru, Malaysia
[3]Advanced Membrane Technology Research Centre (AMTEC), School of Chemical and Energy Engineering, Faculty of Engineering, Universiti Teknologi Malaysia, Skudai, Johor, Malaysia
[4]Department of Materials, Manufacturing and Industrial Engineering, School of Mechanical Engineering, Universiti Teknologi Malaysia, Johor Bahru, Malaysia

Abstract
Membrane technology research has advanced rapidly to treat wastewater, recycle contaminated water, and provide more freshwater. Because of their unique features, such as high porosity (up to 90%) and large specific surface area, electrospun nanofibrous membranes offer a lot of potential for use in membrane processes. When compared to other nanofiber manufacturing techniques, electrospinning can generate new nanofibrous scaffold structures by creating complex assemblies, and it is straightforward to functionalize nanofibers by adding multifunctional elements. Even though there are numerous alternative methods for producing nanofibers, electrospinning is the most versatile. Materials, particularly polymeric nanofibers, have been fabricated by electrospinning, either directly or through

**Corresponding author*: nuhaawang@yahoo.com

Shahid-ul-Islam, Abid Hussain Shalla and Mohammad Shahadat (eds.) Green Chemistry for Sustainable Water Purification, (243–270) © 2023 Scrivener Publishing LLC

postspinning processes. Electrospinning, on the other hand, differs from other nanofiber production processes in that it allows for the creation of a wide range of fiber assemblies. This would undoubtedly improve the performance of nanofiber-based materials while also allowing for application-specific changes. As a result, we must understand the numerous characteristics and methods that allow us to construct the required fiber assemblies. The goal of this book chapter is to discuss the recent development of the produced nanofibrous photocatalytic membranes based on various electrospinning setups, with a focus on physical and morphological structure developments for high wastewater treatment efficiency. The potential of nanofibrous membranes in wastewater treatment, as well as the issues and future approaches, have been comprehensively covered.

Keywords: Electrospinning, photocatalytic membranes, wastewater treatment, polymeric nanofibers

11.1 Introduction

The most promising technology for treating organic contaminated wastewater was revealed to be photocatalytic degradation using nanosized photocatalysts. This is because, in aqueous systems, photocatalysts destroy organic contaminants. Reactive oxygen species (ROS) such as •OH, O_2, H_2O_2, and $1O_2$ are produced by photocatalysts and are responsible for mineralizing organic contaminants and releasing CO_2 and H_2O as innocuous end products [1]. Three components were found to be the most important in the creation of ROS: (1) energy sources such as visible light, ultrasonic, UV light, and heat; (2) major oxidants such as ozone and hydrogen peroxide; and (3) catalysts such as zinc oxide, titania, or other semiconductors. Electrons in the valence band are stimulated to the conductance band when a catalyst is subjected to an energy source (for example, UV radiation). In the valence band, this results in positive charge holes. The electrons and hole pairs then migrate to the photocatalyst's surface to participate in a sequence of redox reactions. When oxygen is present, electrons decrease the photocatalyst before combining with the oxygen to generate superoxide radical anions. Electrons and hole pairs react in water molecules to form ROS. The produced ROS can oxidize organic substances, resulting in innocuous carbon dioxide and water [2].

Tio, ZnO, graphitic carbon nitride, platinum, tungsten oxide, copper oxide, sternum, and other nanosized photocatalysts and composites have recently been produced for wastewater applications. They have a large basic surface area and a diverse functional community, and they are effective photocatalysts. If photocatalytic degradation is to be employed individually in wastewater treatment, several hurdles must be solved successfully, such as difficulty in isolating the suspended photocatalyst from solution and a low reutilization rate [3].

Many investigations have been conducted to immobilize the nanosized photocatalyst in a substrate such as membranes or floating materials to protect the photocatalyst in solution. Immobilization of photocatalysts in porous membranes, commonly known as "photocatalytic membranes." Photocatalytic membranes can reduce photocatalyst loss while retaining effective photocatalytic degradation. PVDF/TiO_2, PVDF/TNT, PSF/Fe-TiO_2, PSF/Cu_2O, and other photocatalytic membranes have been successfully fabricated for wastewater treatment [4]. However, existing photocatalytic membranes need to be improved in terms of photocatalyst dispersion and light use. Traditionally, photocatalytic membranes were made by depositing the photocatalyst on the membrane surface or integrating the photocatalyst into the membrane layer, then casting the membrane. The separation and photocatalytic degradation capacities of these photocatalytic membranes could be limited using this method. On the photocatalyst-coated membrane, the photocatalysts clump together on the membrane surface, limiting the effective catalytic area. Due to poor adhesion, the coated photocatalyst on the membrane surface frequently begins to break off after many reaction cycles [5]. For the photocatalyst-blended membrane, most photocatalysts were integrated into the membrane matrix, although agglomeration within the membrane structure resulted in limited catalyst implementation. Undifferentiated photocatalytic processes can potentially weaken the membrane matrix, reducing the photocatalytic membrane's mechanical durability over time. A modest amount of photocatalyst put to the top surface of the membrane can also diminish photodegradation efficiency. A larger photocatalyst dosage resulted in membrane pore blockage and deterioration of the continuous porous structure, resulting in a significant decrease in membrane permeability due to the formation of agglomerates. As a result, it is critical to look into innovative techniques to cope with this challenging yet critical issue [6].

Nanofibers are made using a variety of procedures, including template synthesis, drawing, self-assembly, phase separation, and electrospinning. It is worth noting that electrospinning is a quick and easy approach to create nanofibrous substrates with a high specific area and porosity, which have a lot of potential in wastewater treatment. Electrospinning is a process for producing nanofibers from a charged or molten polymer solution using a high electric field. In summary, the inorganic precursor, polymer, additive, and volatile solvent in a typical electrospinning precursor solution should all be carefully prepared to ensure that the solution spins swiftly. Changing the electrospinning conditions and the characteristics of the polymer solution can result in a range of morphologies [7]. The most efficient method for manufacturing continuous nanofibers and ultrafine fibers with very small diameters is electrospinning. Small-diameter fibers are preferred because of their stiffness and tensile strength, which can result in improved mechanical performance. Electrospun nanofibrous membranes can also be used to deposit/grow functional nanoparticles for a variety of purposes [8].

In recent decades, the architecture of electrospinning nanofibrous photocatalytic membranes has become more widespread (ENPM). ENPM, which combines a photocatalyst with a nanofibrous membrane, has been developed as a viable option for pollutant degradation in wastewater employing an improved oxidation mechanism. This is due to changes in their physical and morphological structure, which have aided in the improvement of wastewater treatment efficiency. There are more photocatalytic reaction sites available because of the higher specific surface area and porosity, which increases the efficiency of electron separation [9].

Materials, such as polymers and composites, have been fabricated by electrospinning, either directly or through postspinning processes. Electrospinning, on the other hand, is unique among nanofiber synthesis methods in that it may manufacture several fiber assemblies. This would boost the efficacy of photocatalytic membranes made of nanofibers. As a result, we must understand the numerous characteristics and methods that allow us to create the required fiber assemblies. Fiber assemblies can be made from nonwoven fiber mesh, aligned fiber mesh, patterned fiber mesh, random three-dimensional shapes, and submicron spring and tangled fibers [10]. However, more research is needed to fully know and accurately control the dynamics of the production of electrospun fibrous assemblies. For the first time, this research focuses on the design and modification of electrospinning setups to fabricate electrospun nanofibrous membranes with photocatalytic capabilities for wastewater treatment applications. The purpose of this report is to highlight the ENPM's recent accomplishments, with a focus on breakthroughs in electrospinning setup design for high-performance photocatalytic wastewater treatment. This research investigated the effect of designs on the characteristics and photocatalytic effectiveness of ENPM. Furthermore, the performance evaluation, issues, and prospective directions for ENPM application in wastewater treatment are intended to aid researchers in this sector [11].

11.2 Application of Electrospun Nanofibers Polymeric Membranes (ENPM) on Wastewater Treatments

Organic contaminants are organic pollutants and are often composed of biomass, hydrogen, and perhaps other components. Chemical contaminants can cause water turbidity to rise, resulting in eutrophication, which reduces the amount of dissolved oxygen in the system. Chemical contaminants, such as dyes, herbicides, humid chemicals, pharmaceuticals, coal, poisons, and phenolic compounds, are toxic to the setting. Anthropogenic practices, such as domestic, farming, animal husbandry, and industrial drainage, are common sources of organic contaminants in water [12].

Dye-containing wastewater is one of the most important causes of water contamination. Meat, painting, paper, cosmetics, leather, and textile industries all use dyes as a colorant. Traditionally, dyes have been derived from natural plant and insect sources before being quickly converted to synthetic production methods. Unfortunately, certain dyes have been discovered to be mutagenic and poisonous to living organisms. Therefore, pathways to creativity remediation must be studied. Researchers unanimously agreed that developing photocatalysts through electrospinning is the most effective method for improving photocatalytic efficiency. This is because there is more actual surface area and less electron-hole recombination, allowing for faster electron movement [6]. Immobilized 0.02M silver phosphate (Ag_3PO_4) nanoparticles on electrospun PAN nanofibers destroyed MB dye almost completely in 60 minutes, according to Panthi *et al.* As seen in Figure 11.1, Ag_3PO_4 nanoparticles attached strongly to the surface of the nanofiber in agglomerated forms (a). Ag_3PO_4 nanoparticles have been identified as a promising visible-light-driven photocatalyst for the degradation of organic pollutants. This resulted in poor recombination of photogenerated holes (h^*) and electrons (e) [13].

As demonstrated in Figure 11.1, the production of reactive oxygen species is a mechanism for photocatalytic degradation of Ag_3PO_4/PAN composite nanofibers against MB. (b). In the presence of visible light irradiation, Ag_3PO_4 freed electrons from the valence band (VB) and promoted them to the conduction band (CB), leaving photogenerated holes behind. The excited electrons are absorbed by dissolved O_2 in solution, forming $\bullet O_2$ radicals that combine with H^+ to create H_2O_2, forming •OH radicals. The photogenerated holes react with H_20 or OH ions to make •OH radicals at the same time. These radicals are responsible for $AgPO_4$'s photodegradation effectiveness toward MB. The MB depletion process is depicted in

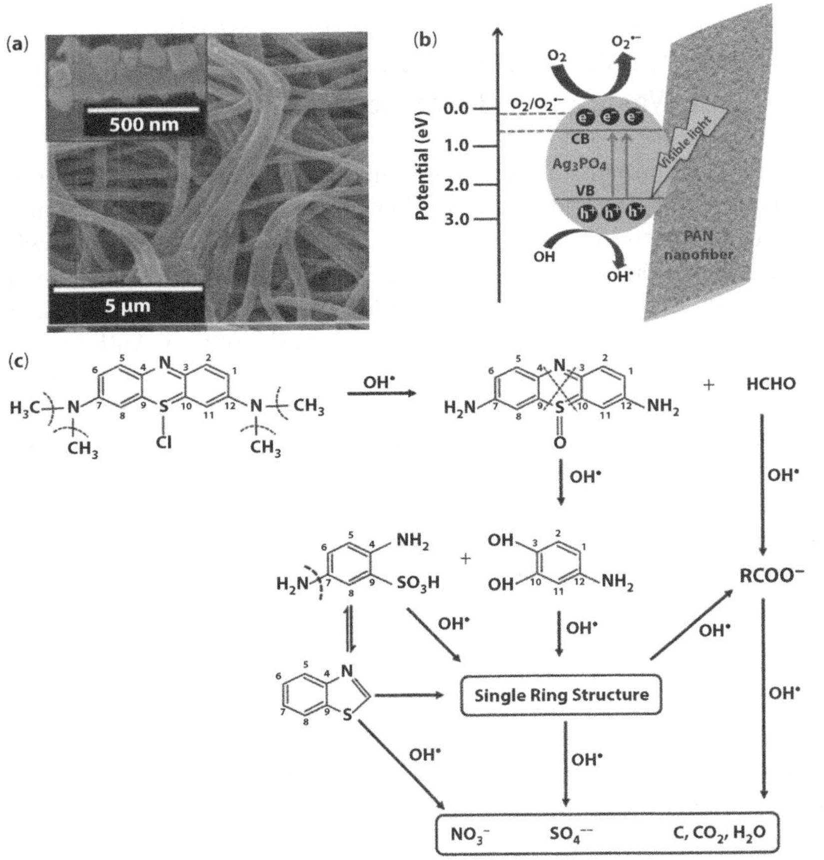

Figure 11.1 (a) TEM image of Ag_3PO_4/PAN nanofiber, (b) schematic diagram of production of radicals by Ag_3PO_4, when exposed to visible light (c) MB photodegradation method as seen in a schematic diagram [6].

Figure 11.1 as an example (c). After being assaulted with •OH radicals, the alkyl groups in N-CH_3 attached to C-12 and C-7 were dissociated and oxidized into HCOH and HCOOH. The bombardment of •OH also disrupts the relationship between C-S and C-N in the surviving structure of MB. Finally, further, simpler organic molecules are created, which are then oxidized until inorganic species, such as Cl^{-1}, H_2O, CO_2, NO_3, and SO^{4-}, are formed [6].

Azo dyes have been prohibited all around the world because of their toxicity and mutagenicity. Azo dyes have vibrant colors such as red, orange, and yellow, as well as the azo property (N=N). Azo dyes include Aniline

Yellow, Bismarck Blue, Congo Red, Direct Black 38, and Methyl Orange (MO). Methyl orange has long been utilized as a model chemical for azo dye photodegradation investigations. Li *et al.* found that tungstosilicic acid hydrate $H_4SiW12O_{40}(SiW_{12})$/cellulose acetate composite nanofibrous membranes degraded MO dye and tetracycline antibiotic better than neat SiW_{12} photocatalyst when exposed to UV light. They discovered that the ideal mass ratio of SiW_{12} to CA was 1:4, with MO and tetracycline degradation efficiency of 94.6% and 63%, respectively. The cellulose acetate nanofibrous membrane's added feature of donating electrons to SiW_4 during MO degradation could explain why MO decayed more quickly than TC under the same conditions. This interaction has hastened the breakdown of MO rather than tetracycline. Incorporating SiW_{12} photocatalyst into an electrospun cellulose acetate substrate further increased efficiency by providing a more efficient and larger contact surface between the pollutant and the SiW_{12} photocatalyst [14].

Qayum *et al.* investigated the photocatalytic activity of PAN/AgBr/Ag fiber membrane against salicylic acid. The barrier photodegraded salicylic acid in under 5 hours, the researchers claim. They further stated that the PAN/AgBr/Ag membrane was stable after five cycles, with up to 96% degradation efficiency. Lian and Meng successfully discolored MB dye solution using electrospun TiO_2/bioglass hybrid nanofiber [15]. Efficacy of TiO_2 wallpaper membrane and TiO_2 film photocatalysis compared. The photocatalytic efficiency of the membrane was studied by altering the top layer of TiO_2 wallpaper thickness. After 120 minutes, the thickest top layer (0.051 mm) membrane had 60% MB discoloration, compared to 33.5% for the TiO_2 film. This preserves the nanofibrous' porous structure while increasing the photocatalyst's surface area. So, the discoloration reaction may be more effective. The porous structure of the TiO2 film may be closed as the layer of TiO_2 film is raised to four layers, reducing the photocatalytic output. Since heat can bond multilayered photocatalyst wallpaper to non-photocatalyst carriers [15].

As a photocatalyst for organic pollutant degradation, polyoxometalates (POM) have recently gained attention. POMs are transition metal-oxygen clusters that are nontoxic, adjustable in size, and photostable. POMs are water and organic solvent-soluble, making recovery and reuse difficult. POMs also have a tiny surface area. So genuine POM host materials are urgently required to allow recovery and increase contact area for practical applications. One of the most studied POMs was silicotungstic acid, $H_4SiW_{12}O_{40}$ (SiW12). Its photocatalytic activity and chemical stability are due to its Keggin structure (W-O-W). Li *et al.* (2017) investigated the effects of different mass ratios of SiW12 to cellulose acetate substrate (1:5, 1:4, and 1:3) on ENPM tetracycline and methylene blue degradation [13].

Cellulose acetate was chosen for its biodegradability, hydrophilicity, high mechanical qualities, and chemical resilience. The SiW_{12} to CA 1:4 membrane was shown to have a greater rate of TC and MO degradation, with 63.8% and 94.6%, respectively. The photocatalytic performance declined as the SiW_{12} loading in CA was raised further, which was attributable to the formation of beads in the membrane, which reduced the photocatalytic membrane's basic surface area. Figure 11.2(a–d) shows SEM representations of the manufactured membranes, with the evolution of beads arrangement in membrane 1:3 visible rather than membrane 1:4. In addition, as the mass ratio of SiW12 to CA approaches 1:3, the hydrophilicity of the membrane diminishes, as seen in Figure 11.2(e). This could inhibit complete interaction between the photocatalyst and impurities, reducing the photocatalytic capacity of the membranes [13].

The global expansion of agriculture has required the widespread use of pesticides to control unwanted vegetation like weeds and grass. Herbicides can be absorbed into the soil before entering the water table. Humans and animals ingest herbicides present in industrial wastewater. Xie *et al.* claim ENPM can degrade colors and low toxicity phenyl urea herbicides like isoproterenol. Figure 11.3: Isoproterenol degradation by CQDs-$Bi_{20}TiO_{32}$/PAN electrospun fiber membranes. Typical isoproterenol elimination steps were as follows: (1) The fiber surface CQDs-Bi20TiO_{32} photocatalysts absorbed visible light energy to create electrons in the conduction band (ECB) and photon in the valence band (hVB+) [16].

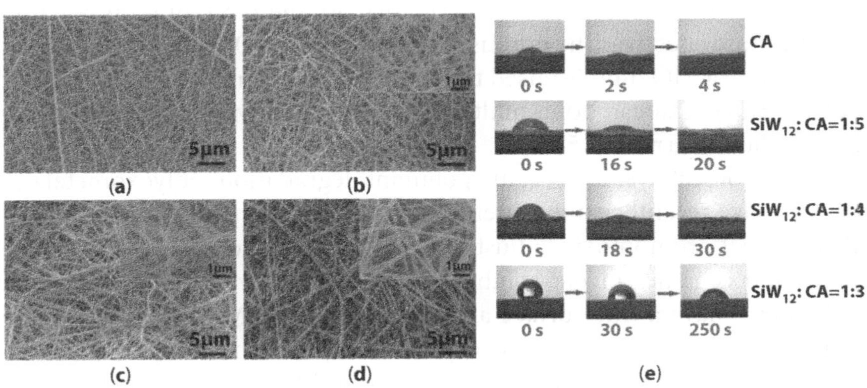

Figure 11.2 SEM images of (a) neat CA nanofiber membrane (b) SiW_{12}/CA composite nanofibers with ratio of SiW_{12}:CA = 1.5, (c) SiW_{12}:CA = 1.4, (d) SiW_{12}:CA = 1.3, (e) water contact angle of the fabricated membranes [13].

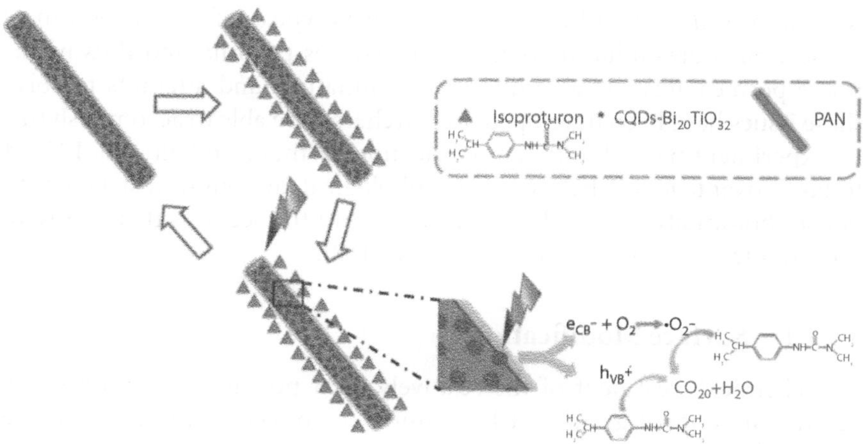

Figure 11.3 Degradation of isoproterenol mechanism by CQDs-$Bi_{20}TiO_{32}$/PAN electrospun fiber membranes [16].

11.3 Improvements in Morphology and Physical Structure of ENPM

ENPMs now have better physical and morphological features for photocatalysis and filtering. Several variables explain this pattern. In addition to increasing the number of photocatalytic reaction sites, they also improve electron-hole pair separation efficiency. Second, the unique electrospinning approach allows for fine-tuning surface functionality during processing, efficiently dispersing photocatalyst into a polymer-based nanofiber. This state promotes full source light irradiation of the photocatalyst, increasing photocatalytic efficiency. Third, the separation process is tripled due to the 3-D interlaced pore structure of electrospun nanofiber membranes and their superior porosity compared to typical polymeric membranes with a 2-D framework. When removing contaminants from feed, a multitarget separation system is preferred. Finally, compared to normal phase inversion membranes (5–35%), ENPMs offer excellent pore interconnectivity, constant pore sizes, and porosity of over 80%. These features improve water filtration efficiency by increasing membrane permeability [17].

ENPM has attracted the interest of many people in this regard due to its enormous ability for decomposing organic molecules in wastewater. Nevertheless, one typical issue with nanofibrous materials is that increasing the working pressure induces pore structural deformation and a reduction in inflow. In hydraulic flow, this is typical of electrospun mats because pressure-induced compression reduces porosity and thus membrane permeability. As a result, several initiatives and attempts to solve these issues have been developed. Researchers were able to accomplish this by experimenting with different fabrication methods and altering ENPM to take advantage of advances in morphological and physical structure to boost photocatalytic and filtering efficiency. In the sections that followed, the specifics were examined and reviewed [18].

11.3.1 Surface Modification

Wet chemical treatment of the relatively inert polymer nanofiber yields reactive functional groups such as amine, carboxyl, hydroxyl, or nitrile groups. On the polymer molecular level, the nuclear or electrophilic attack is feasible. This is usually done in an alkaline medium. The polymeric nanofibrous membrane substrate surface can be changed to enhance photocatalyst particle interaction or deposition.

Panthi *et al.*, for example, reported electrospinning Ag_3PO_4/PAN composite nanofibers and subsequently surface modification of PAN nanofibers by treating with NH_2OH aqueous solution for 15 minutes at 65°C [19]. During the treatment, PAN's nitrile groups were converted to amidoxime groups, which were employed to anchor Ag^+ ions on the surface of PAN nanofibers, as seen in Figure 11.4. The Ag^+ ions are then combined with the PO_4^{3-} ions to form Ag_3PO_4 nanoparticles. The system's use of amidoxime functional groups in combination with Ag_3PO_4 nanoparticles was thought to be an effective strategy for correct nanoparticle attachment, resulting in greater efficiency.

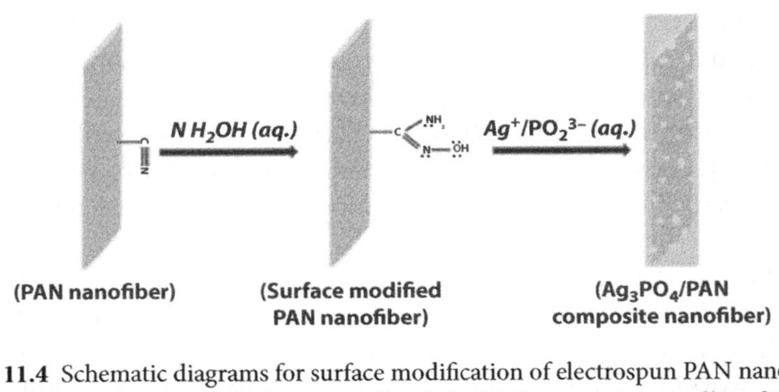

Figure 11.4 Schematic diagrams for surface modification of electrospun PAN nanofibers with amidoxime groups and development of Ag_3PO_4/PAN composite nanofibers [19].

Kim *et al.* claim that the immersion process allows for polydopamine functionalization of nanofibers, changing their surface and facilitating particle interaction. Photocatalytic nanoparticles may now be embedded in nanofiber surfaces while retaining their mechanical properties. An aqueous ZnO nanoparticle solution was used to bind the photocatalyst to the polydopamine-coated PU nanofibrous mat. The surface of the nanofibers was then hydrothermally generated with ZnO nanorods. Figure 11.5 shows the fabrication process. Hydrothermal treatment was proposed to increase photocatalyst nanoparticle adhesion within the electrospun nanofibrous polymeric substrate. Kim *et al.* developed Pdopa-ZNRs/PU ENPM using electrospinning, surface functionalization, and hydrothermal treatment. 12 hours in a dopamine hydrochloride solution with tris-HCl buffer [20].

The ZnO nanorods adhered to the polydopamine functionalized nanofibers, the researchers claim. The polydopamine-ZNRs/PU nanofibers had strong photocatalytic/antimicrobial activity and reusability at low-intensity UV-LEDs. Polydopamine's catechol group acts as an electron trap, lowering e–h pair recombination, and increasing ENPM photocatalytic efficiency [20].

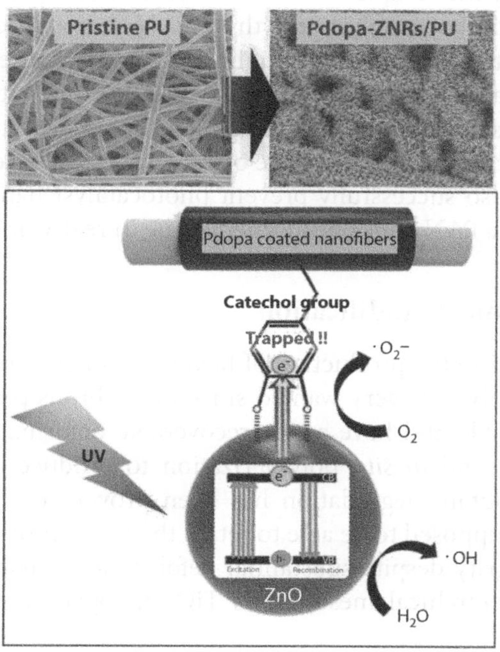

Figure 11.5 Schematic flow path of fabrication of Pdopa-ZNRs/PU through surface functionalization and hydrothermal treatment [3].

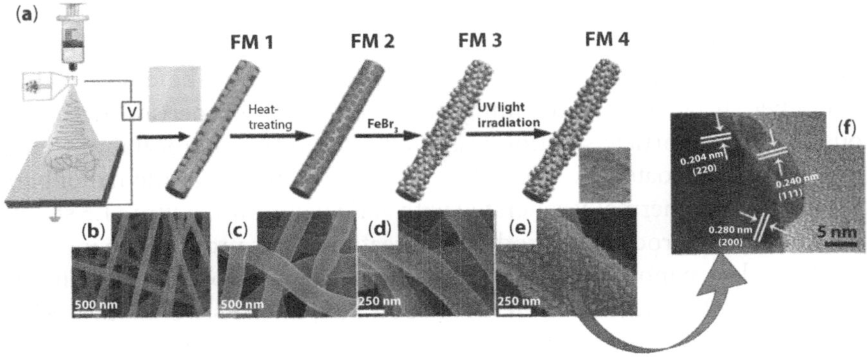

Figure 11.6 (a) The four steps of the PAN/AgBr/Ag fibrous membrane fabrication technique are depicted in this diagram, (b–e) (f) HRTEM image of the AgBr/Ag on the membrane surface, SEM images of the fabricated membranes from each stage [15].

A multifunctional polyacrylonitrile (PAN)/AgBr/Ag fibrous membrane was designed and manufactured by Qayum *et al.* [21]. Initially, the dope remedy failed. They produced a PAN/AgBr/Ag fibrous membrane in four phases, labeling the samples FM1, FM2, FM3, and FM4 (a). (b–e) SEM pictures showed photocatalyst growth on PAN fiber after electrospinning, heating, FeBr3 treatment, and UV light irradiation. Figure 11.6 shows the growth of Ag nanoparticles on the surface of AgBr implanted in PAN fiber (f). The findings indicated that creating an AgBr/Ag Schottky circuit could increase photocatalytic performance by increasing AgBr's electron density [21]. It could also successfully prevent photocatalyst nanoparticles from peeling from the PAN nanofibrous membrane in real-world applications.

11.3.2 Chemical Modification

A novel method to the production of flexible, stronger, and stable nanofibrous photocatalysts is very wanted since nanofibrous photocatalysts are typically breakable and have a tiny recoverable deformation. The use of electrospinning and *in situ* polymerization to produce inorganic-based ENPM for pollutant degradation has been proven to be effective. This approach was supposed to be able to retain the nanofibrous photocatalyst's structural integrity despite substantial deformation. Zhang *et al.* created flexible and hierarchical mesoporous TiO_2 nanoparticles modified TNF

(TiNFNP) composites by combining electrospinning and *in situ* polymerization [4].

This technique uses pristine electrospun titanium dioxide nanofibers (TNF) as a template to decorate TiO_2 photocatalyst nanoparticles. The white electro spun TiNF was then immersed in a solution containing TiO_2 nanoparticles, bifunctional benzoxazine, and acetone before being dried and processed with in situ polymerizations of BA-a monomers at 200°C for 1 hour, resulting in the formation of a Poly benzoxazine (PBZ) layer on the TNF surface. The color of the created PBZ/TiNF was yellow. PBZ was utilized as a carrier and fixative to prevent photocatalyst nanoparticle aggregation. After that, the membrane was calcined at 550°C for 2 hours under N_2 flow, yielding a black TiO_2 nanofibrous membrane. A schematic diagram of the fabrication process, as well as the membranes that are produced, is shown in Figure 11.7 [4].

The MB deterioration of the constructed ENPM was evaluated, and the MB was shown to deteriorate after 30 minutes. In terms of photodegradation, the manufactured membrane surpasses commercial catalysts like P25. Importantly, the rapid synthesis of these intriguing materials could pave the way for photoreactors based on the nanofibrous membrane to remediate water and air pollutants.

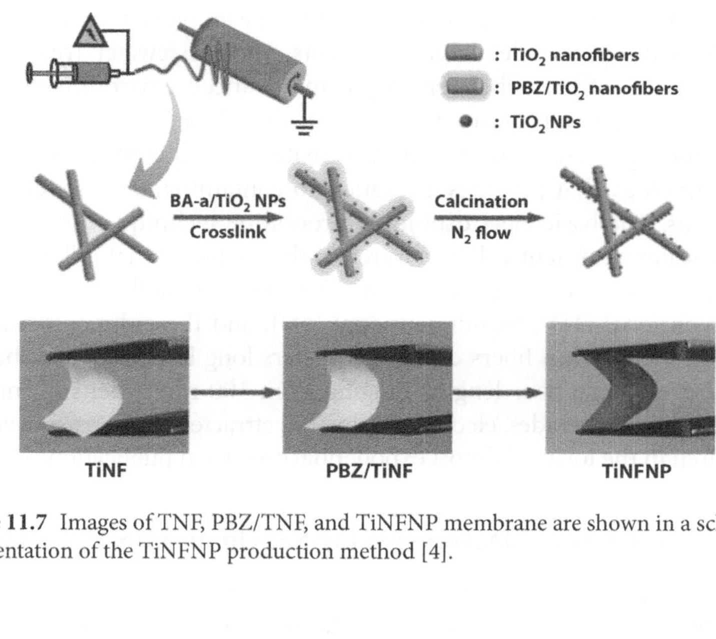

Figure 11.7 Images of TNF, PBZ/TNF, and TiNFNP membrane are shown in a schematic representation of the TiNFNP production method [4].

11.4 Setup and Configurations of Electrospinning for Core-Sheath Structures of EPNM for Photocatalytic Membranes

Electrospun polymer nanofibers have been the subject of much research to understand the fundamental phenomena of the manufacturing process as well as the physical and chemical properties from a material science approach. Researchers will be able to better monitor the fiber component, morphology, and aggregate structure, as a result, offering electrospun nanofiber applications a significant boost. Electrospinning nanofibers, in particular, have shown significant evidence of potential roles in resolving energy and environmental concerns, which are two of the most pressing issues confronting society in the 21st century [14].

In emerging countries, the need for clean water and air is growing, and it will be a major stressor for modern economies in the future. To address this environmental concern, nanofibrous membranes have been widely explored for usage in sensors, photocatalytic, air filtration, liquid filtration, oil spill cleaning, self-cleaning, adsorbents, and electrical insulation [18]. Sensors are a significant application of nanofibers in the linked sectors, including the many forms of nanofiber sensors, to multiple analytes in the air or water since environmental control is a required necessity for environmental management. As a result, useable nanofibers could be employed for air filtration, oil spill clean-up, self-cleaning, adsorbents, and wastewater treatment. A cursory review of nanofibers employed in advanced environmental applications over the last 10 years demonstrates that their impact is significant, and promising electrospun nanofibrous materials will continue to reflect a vital technology that provides a solution to contaminants in water and air ecosystems. The basic electrospinning processing technique for nanofibers in wastewater treatment will be covered in the subtopics that follow.

Electrostatic drawing (electrospinning) is another method of generating one-dimensional (1D), two-dimensional (2D), and three-dimensional (3D) materials. Electrospun fibers can be kilometers long because nanotubes and nanotubes generate fiber lengths ranging from 100 nm to tens of microns. Over the last two decades, electrospinning has attracted an increasing volume of research in the form of almost exponential growth in publications [6].

11.4.1 Impacts of Electrospinning Set Up on EPNM Structures

Corporate involvement in different sectors of the market, such as air and water filtration systems routed by pharmacology problem innovation

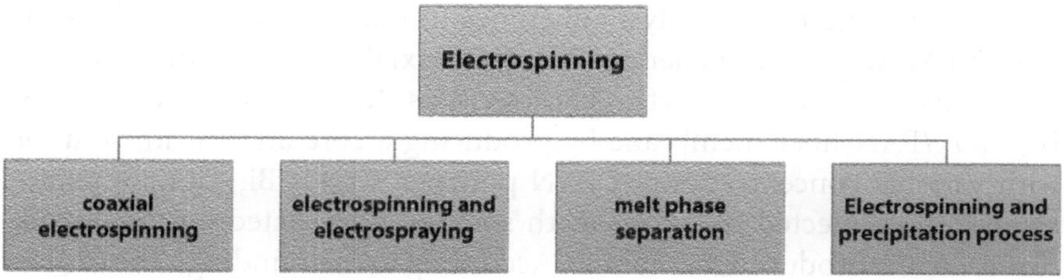

Figure 11.8 Electrospinning and fabrication methods.

and regenerative medicine (TE & RM), as well as industrial areas based on nanomicron and submicron fibers, with diverse nonbiological content and physicochemical characteristics, has been recognized as field transformation. Technology advancements in the so-called 1D characteristics of ultrafine fiber, where positively directed structural elements along the length of the fiber firmly restrict material and electronic diffusion distances perpendicular to the fiber axis, could support energy conservation, generation, and electronics. The classes of electrospinning processing methods are illustrated in Figure 11.8.

These materials can currently be manufactured and synthesized using several methods. Since these technologies allow for the development of 1D organic nanostructures and polymer fibers on a laboratory scale, electrospinning blends unrivaled laboratory operational simplicity with industrial production capacity. The sections that follow will go into every aspect of electrospinning.

11.4.1.1 Coaxial Electrospinning

Around the year 2003, coaxial electrospinning technology was developed to create core-shell tailored fibers with a variety of material qualities. By combining two distinct types of polymers, the coaxial electrospinning technique can be utilized to create core-shell structured micro nanometer-scale threads. The use of coaxial electrospinning to make core-shell structured fibers has grown in popularity to improve material property profiles for possible applications, particularly in environmental purification. The coaxial electrospinning process appears to be successful in preventing core and shell polymer mixing, as well as contaminating chemicals. This approach also has good mechanical characteristics, porosity, and hydrophilicity, as well as the ability to keep membranes from bulging. This is due to the synergistic effect of combining two distinct polymers, in which the strength of the other polymer overcomes the weakness of the polymer [22].

By permitting photocatalyst adherence to the fiber surface to be firm, the ENPM, which was generated via the coaxial electrospinning process, maintains good photocatalytic characteristics. Xie *et al.* created a CQDs-$Bi_{20}TiO_{32}$/PAN fiber membrane by producing a core and sheath solution with variable concentrations of PAN polymer. CQDs-$Bi_{20}TiO_{32}$, a photocatalyst, was injected into the sheath solution and coated on the outside surface of the produced membrane. According to their findings, the CQDs-$Bi_{20}TiO_{32}$ photocatalysts were equally anchored to the composite fiber surfaces. This state exposes the photocatalyst to source light, which promotes photocatalytic performance. The average fiber diameter expanded as the photocatalyst loading increased, and the fiber surface got rougher. This could be because the sheath solution has a higher viscosity, preventing the fiber from expanding during the electrospinning process [16].

The ENPM was made utilizing a coaxial electrospinning approach, which improves the morphological structure by creating a new nanofibrous photocatalytic membrane architecture. Tang *et al.* recently reported that using a coaxial electrospinning technique, they were able to successfully produce a unique TiO_2 microsphere/nanosphere structure on the surface of nanofiber. The core solution contains polyvinyl acetate, butyl titanate, acetic acid, and tert-butyl alcohol. The shell solution is made

Figure 11.9 SEM image of nanofibers embedded in TiO_2 micro/nanospheres: (a) a beaded structure and (b) a beaded structure (b) embedded structure.

up of ethanol, polyvinylpyrrolidone, acetic acid, and butyl titanate [23]. Both solutions were pumped into the core and shell micropumps of the electrospinning equipment. The SEM images obtained using this method revealed two manufactured membrane structures, a bead structure, and an embossed structure, both of which had been calcined at 550°C (Figure 11.9). The spherical products were not autonomously expelled but securely embedded on the nanofiber's surface after drying and calcining, and the electrical field strength was used to form the patterns [23].

11.4.1.2 Electrospinning and Electrospraying

The combination of electrospinning and electrospraying is another interesting technology to produce polymeric-based ENPM that caught the researchers' attention. Using electrical forces, the electrospraying process atomizes liquid material. Coulombic repulsion between the surface charges disperses the dope solution ejected from the needle nozzle into finely charged droplets when the electrical potential is high. The feed rate and voltage settings can influence the size of the droplet. The previous study has indicated that combining electrospinning and electrospraying in the fabrication of flexible photoelectrodes for solar cells and fuel cells is advantageous. As a result, combining these two methodologies may lead to the development of an ENPM that is both adaptable and effective [24].

Ramasundaram *et al.* created highly adaptable and porous poly (vinylidene fluoride)/TiO_2 membranes for photocatalytic applications using successive electrospinning and electrospraying processes. On the surface of the electrospun PVDF mat, the photocatalyst dispersion of TiO_2 was electrosprayed (Figure 11.10). According to the morphological examination, the electro sprayed TiO_2 is deposited as clusters on the PVDF mat in a single pot. The photocatalytic behavior of the electro sprayed TiO_2/PVDF mat as compared to the nonporous PVFF/TiO_2 film made by solution casting with a similar amount of photocatalyst. The photocatalytic activity of the PVDF/TiO_2 mat was substantially higher than that of the PVDF/TiO_2 film. The porous structure of the electrospun PVDF/TiO_2 mat allows for greater photocatalyst activation sites, resulting in increased photocatalytic efficiency. The creation of photocatalyst clusters on the upper surface of the PVDF mat, on the other hand, will accelerate separation from the substrate during actual operation. Furthermore, this technology involves two different electrospinning and electrospraying procedures, each of which adds expenses and time to the process [25].

The SEE system can be formed by combining electrospinning and electrospraying, which are fundamentally the same electrohydrodynamic-based

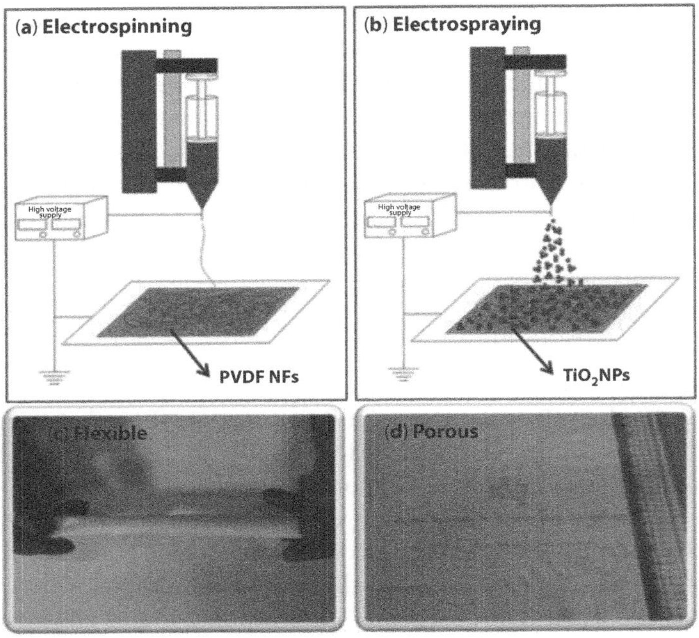

Figure 11.10 Diagram of a setup for electrospinning and electrospraying simultaneously period [20].

processes. Pahasupan created and interlocked tiny TiO_2 photocatalyst clusters within a nonwoven nylon-6 electrospun membrane network using the SEE process. Anatase-TiO_2 suspensions were prepared at concentrations of 2.5%, 5%, and 10% by weight. The nylon-6 solution and the TiO_2 suspension were each put into their syringe pumps and ejected through a metallic needle attached to an opposite-polarity high voltage source. The membrane's base layer was made by electrospinning nylon-6 solution at a voltage of 25 kV for 15 minutes. The nylon-6 nanofibrous network and TiO_2 suspension at concentrations of 2.5%, 5%, and 10% (TN-2.5, TN-5, and TN-10, respectively) were electrospun and electro sprayed simultaneously for 4 hours before drying in an oven at 80°C for 24 hours to generate a nanofibrous photocatalytic membrane [3].

As evidenced by the related XRD spectra, the range of the produced membranes is closely aligned with the anatase-TiO_2 concentration, suggesting that the TiO_2 nanoparticles were successfully integrated into the nylon-6 membrane utilizing SEE. The UV-VIS transmittance spectra demonstrated that the production of TiO_2 microclusters in the PEVM generated using this process did not affect the optical characteristics of anatase-TiO_2 particles with virtually identical photocatalytic activity within the same wavelength range of 390 to 393 nm [3].

All Nylon-6/TiO_2 PEVM produced using SEE technology achieved a removal rate of about 99% after 120 minutes using MB as a model pollutant. These findings suggest that using simultaneous electrospinning and electrospraying to reduce nanoparticle photocatalyst separation from the nanocomposite membrane is an effective and easy strategy. The heaviest loaded membrane, the TN-10, however, lost roughly 10% of its weight after 5 cycles of photocatalytic therapy due to photocatalyst micro cluster breakdown, as seen in Figure 11.11. The mechanical strength of the Nylon-6/TiO_2 ENPMs in their current state was moderate, according to the findings. Further advancements in the mechanical stability of the ENPM manufactured by EES are necessary to resist the rigorous circumstances of real-world water treatment [26].

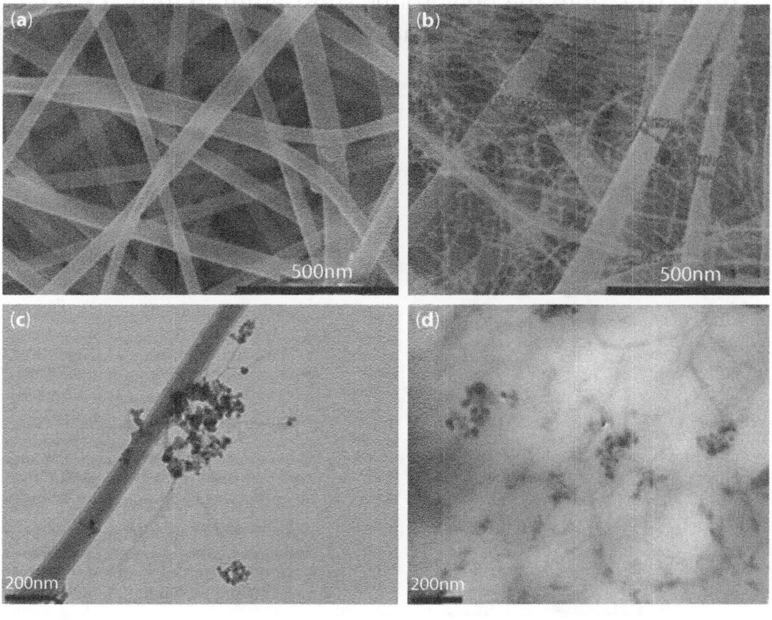

Figure 11.11 FSEM images of the pristine nylon-6 mat (a) and TiO_2/nylon-6 mat (b). TEM images of TiO_2/nylon-6 nanofibers (c) and microtome TEM of TiO_2/nylon-6 nanofibers (d).

11.4.1.3 Separation of the Melt Phase Technique

A simple process known as suspension coating and melt-phase separation was recently used to fabricate large-scale photocatalytic nanofibrous membranes. Previous research on nanofibrous photocatalytic membranes tended to focus on the individual performance of photocatalytic and filtration processes, with only a few studies focusing on the two processes' synergistic performance. This is because of its low mechanical stability, low reusability, and unusually high porosity, all of which might lead to weak photocatalyst-membrane support contact. As a result of the melt phase separation procedure, the morphological and physical structure of the membrane may be improved, which are important factors for membrane filtration efficacy [27].

The TiO_2 photocatalyst was distributed before using the melt extrusion process, PVA-co-PE was mixed with cellulose acetate butyrate (CAB), resulting in mixed composite pellets, as illustrated in Figure 11.12 (a). Temperatures ranged from 190 to 230 ° C throughout processing. Another study employing a similar approach sprayed the composite solution CAB/PVA-co-PE/TiO_2 onto the polyproline substratum to form a nanofibrous membrane for photocatalytic degradation and filtration of MB [28]. Three different concentrations of TiO2 were utilized to boost photocatalytic activity and filtration effectiveness: 1%, 3%, and 5%. They discovered that

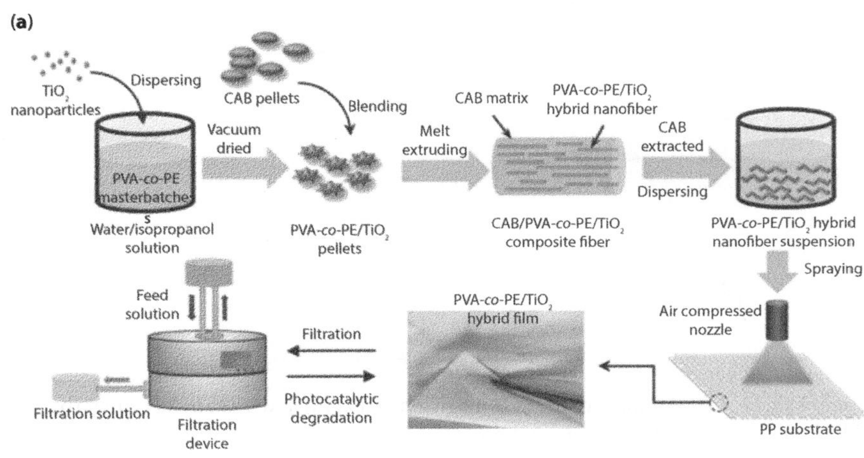

Figure 11.12 (a) The fabrication of a PVA-co-PE/TiO_2 hybrid membrane and the filtration procedure is depicted in this diagram [28].

increasing the TiO_2 loading did not influence the synthesized membranes' water contact angles.

11.4.1.4 Process of Electrospinning and Precipitation

Various ENPMs have been created in recent studies using a mix of electrospinning and UV calcination. There was also a paper there that discussed the creation of ENPM using in situ polymerization. PMs with photocatalysts packed in nanofibers are more likely to exhibit intriguing stability than photocatalysts on top of nanofibers. The dispersion problem of precursor solution mixing, however, is a key challenge for ENPM processing. As a result, a variety of approaches were investigated as possible candidates for creating ENPM to effectively treat contaminants in wastewater [29]. Electrospinning and precipitation techniques are combined in one of them. For example, Bai *et al.* created a unique stand-alone p-n nano heterojunction photocatalyst with YSZ nanofibers as the backbone to support TiO_2 NPs and Ag_2O nanocrystals, as illustrated in Figure 11.13, electrospinning and precipitation were used in combination.

The backbone of the YSZ-TiO_2 nanofibers was created by electrospinning and subsequently calcined at 800 degrees Celsius. The Ag_2O@YSZ-TiO_2 p-n nano heterojunction ENPM was precipitated by immersing the produced YSZ-TiO_2 membranes in alkaline $AgNO_3$ solution and exposing them to a 24 W UV light for 1 hour. Adjusting the weight ratio of $AgNO_3$ and YSZ-TiO_2 membranes from 0.1:1, 0.3:1, and 0.5:1 to 0.5:1 yielded AZT-1, AZT-3, and AZT-5. The best weight ratio of AgNO3 and YSZ-TiO_2

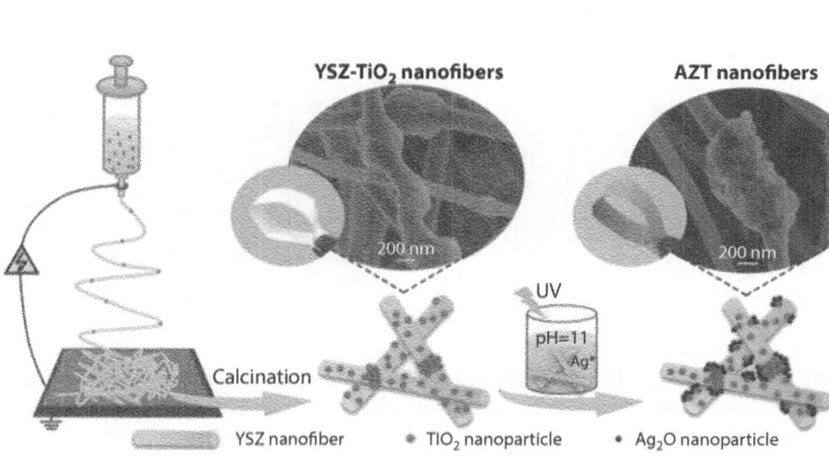

Figure 11.13 Using a combination of electrospinning and precipitation methods, a flexible Ag_2O@YSZ-TiO_2 ENPM can be produced [5].

according to the Kubelka-Munk equation was 0.3:1 (AZT-3), which had the lowest bandgap energy of 2.89 eV [5].

The results of photocatalytic decolorization of MB aqueous solution under UV irradiation were compared, and the findings are shown in Figure 11.14 (a). The concentration of MB was unaffected by UV irradiation in the blank experiment without photocatalyst. The YSZ nanofibrous membranes show no photocatalytic performance and are even similar to the blank experiment due to the high bandgap (5.0 eV) of zirconia. Surprisingly, after the anatase TiO_2 NPs were injected into the YSZ nanofibers, the photocatalytic performance of the YSZ-TiO_2 membranes improved considerably. After the surface texturing of Ag_2O nanocrystals on the YSZ-TiO_2 nanofibers, which was induced by the Ag_2O/TiO_2 p-n nano heterojunction in the AZT fibers, the photocatalytic action was greatly enhanced.

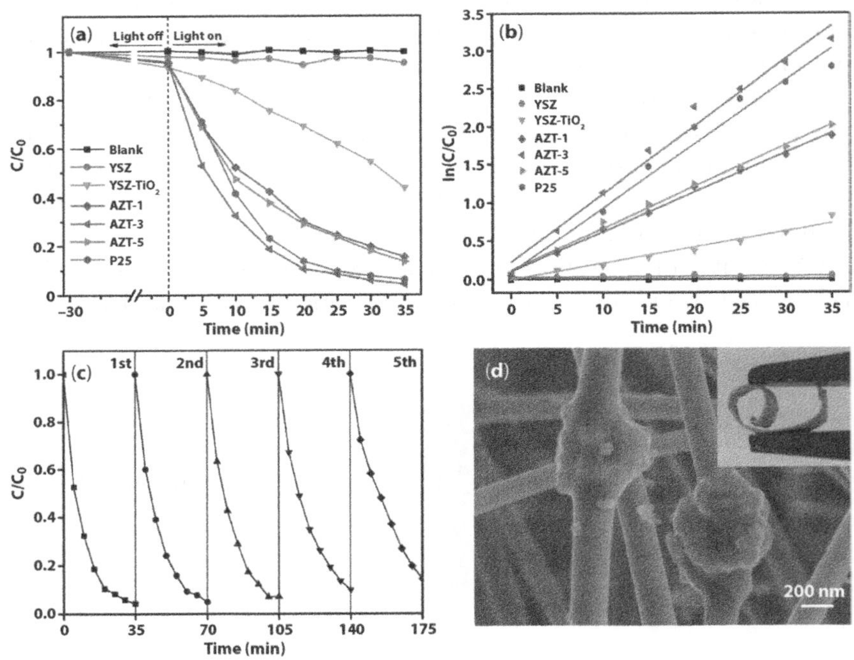

Figure 11.14 Under UV light irradiation, (a) MB photocatalytic degradation profiles and (b) Kinetic linear fitting curves of MB degradation over YSZ, YSZ-TiO2, AZT-1, AZT-3, AZT-5 nanofibrous membranes, and P25 (c) For 5 cycles, photocatalytic activity of AZT-3 nanofibrous membranes for MB breakdown. (d) After five cycles, FE-SEM picture of the relevant membranes. The related shot inset shows the membranes' flexibility after five cycles [5].

11.5 Future Directions and Challenges

The general opinion is that electrospun nanofiber photocatalytic membranes are fragile, brittle, and prone to breaking into individual pieces. As a result, ENPM is frequently used as a mat that is simply soaked in unclean water. However, to extract the mat from the treated water, this technique demands additional treatment. The deposition of ENPM in filtered water resulted in secondary waste generation. PM should be included on a sturdy and durable substrate to maximize the mechanical strength and durability of the electrospinning nanofibrous photocatalytic membrane. At the same time, it aided in the processing and recycling of waste.

According to the research, the bulk of studies showed that the manufactured electrospinning nanofibrous photocatalytic membranes were particularly submerged in polluted solutions for photocatalytic degradation, but the membrane's key characteristic, filtering, was ignored. As a result of this scenario, the photocatalytic degradation effectiveness of the produced nanofibrous photocatalytic membrane is reduced. The importance of filtration in photocatalytic destruction of pollutants by nanofibrous photocatalytic membranes, as well as its contribution (CITE) has been reported. Furthermore, the mechanism behind photocatalytic degradation and filtration coupling is unknown. There is also a scientific question that requires more research, such as why filtration can improve photocatalytic degradation efficacy. As a result, comprehensive research into this mechanism is required, which will help to guide future research and applications of photocatalytic membrane materials.

ENPM was created using solution-based electrospinning in general. Photocatalyst nanoparticles may leak from the substrate because of this approach. Melt electrospinning, on the other hand, is an intriguing approach for manufacturing ENPM. Melt electrospinning was first proposed in Norton in 1936 and developed fully in the 20th century. Melt materials by electrospinning them with hot air, an oil pan, and an electrical laser. This process is effective for making stiff, insoluble polymers like polytetrafluoroethylene, polyethylene, and polypropylene that are difficult to dissolve in a solvent. Aside from that, melt electrospinning has the advantage of being a safer and more environmentally friendly technique because harmful solvents may be avoided. As a result, the residual solvent that is generally utilized in the solution-based electrospinning process does not need to be treated. Figure 11.15 shows a visual example and schematic diagram of a melt electrospinning apparatus with a laser as a heating source.

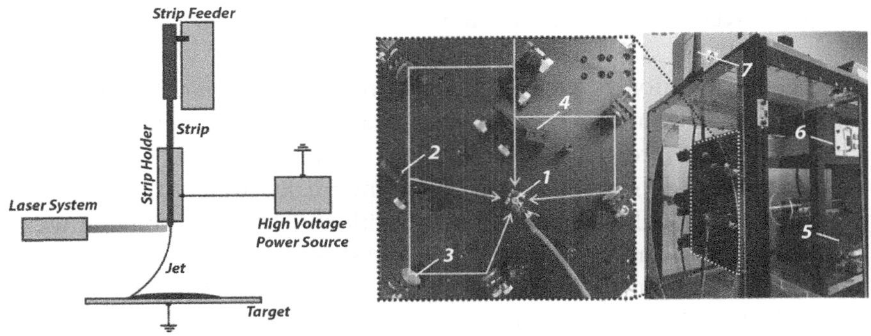

Figure 11.15 Schematic diagram and a digital image of laser-heated melt electrospinning setup (labeled 1: nozzle, 2: beam splitter, 3: reflector, 4: beam baffle, 5: feeder, 6: high voltage system, 7: Melt electrospinning is a method of producing fibrous structures from polymer melts. The collection of the fibre can be very focused in this case of electrospinning. Melt electrospinning writing, when combined with moving collectors, is thus a method of performing 3D printing with ultrathin fibres) [30].

The melt electrospinning rig is made up of four parts: A feeding device (nozzle and feeder), a high-voltage system, a CO_2 laser system (beam splitter, reflector, beam baffle, laser transmitter), and a grounded collecting target are all required. The molten membrane is frequently structured as a larger fiber in this process due to the higher viscosity, making it less desirable than the solution-based electrospinning procedure, which is vulnerable to the formation of the micron-sized fiber. In terms of immobilization of photocatalyst nanoparticles into the polymeric substrate, melt electrospinning produced an improvement. As a result, these proposed remedies were designed to address ENPM's primary flaws in actual wastewater treatment, including basic limits like solvent recovery, increased voltage utilization, and lower mechanical stability.

High voltage consumption, high electric field dependence, and low throughput are all key constraints for ENPM manufacture. As a result, it is critical to improve the nanofibrous photocatalytic membrane production technology now in use. Solution blow spinning, plasma-induced synthesis, centrifugal spinning, and melting electrospinning are only a few of the new nanofibrous photocatalytic membrane solutions that might be offered. Solution blow spinning, which uses compressed air to shape the fibers, can be used to create the nanofibrous photocatalytic membrane.

Solution blow spinning has the potential to solve the drawbacks of traditional electrospinning, such as the high electrical voltage required and the difficulties of fabricating nanofibers *in situ*. This approach combined electrospinning and melt-blowing processes to achieve a greater output volume. Another scalable method for producing ENPM is plasma-induced synthesis. With the help of the plasma generated between the electrodes, the ENPM may be formed utilizing this technique. The plasma would be created without the use of any chemicals or gases underwater. By rapidly blasting electrode surfaces with extremely energetic radicals, the ENPM might be created. Centrifugal spinning, rather than using an electrical field, might be presented as a method of producing ENPM by utilizing centrifugal force to accomplish the promised production of nanofibrous membranes. The nanofiber is extruded using the centrifugal force of a spinning reservoir. In general, the controlled monitoring of centrifugal force, viscoelasticity, and mass transfer features of spinning solutions is used to deplete polymer jets into nanofibers. Centrifugal force was projected to be 500 times more effective than standard electrospinning in the creation of nanofibrous membranes.

11.6 Conclusion

Major efforts to solve the problem of water and energy scarcity have been motivated by the growing necessity to do so. Nanotechnology is becoming increasingly popular in the field of water treatment around the world. The fascinating features of nanomaterials, in combination with present technology, hint at a bright future for wastewater treatment revolutionization. For a long time, scientists and engineers have been attempting to combine photocatalytic and nanofiber technologies to clean wastewater. The current state and prospective improvements in photocatalytic nanofibrous technology are discussed in this paper, with an emphasis on their synthesis, structural and morphological enhancement, and water treatment applications. This research looked into modern procedures like electrospinning, coaxial electrospinning, electrospraying, melt phase separation, and precipitation. Surface modification, in situ polymerization, calcination, drying, hydrothermal, and carbonization are all chemical and thermal processes for posttreatment of ENPM. The deficiencies of ENPM science advancement, as well as prospective discoveries, were thoroughly examined to realize their future practical applications. This study adds to our understanding of modular ENPM for wastewater treatment.

11.7 Acknowledgment

The authors are thankful to Universiti Kuala Lumpur Malaysian Institute of Industrial Technology (UniKL MITEC), Center of Research and Innovation (CoRI) UniKL.

References

1. Awang, N., Ismail, A.F., Jaafar, J., Matsuura, T., Junoh, H., Othman, M.H.D., Rahman, M.A., Functionalization of polymeric materials as a high-performance membrane for direct methanol fuel cell: A review. *React. Funct. Polym.*, 86, 248–258, 2015.
2. Zhang, W., Zhang, Y., Lu, C., Deng, Y., Aerogels from crosslinked cellulose nano/micro-fibrils and their fast shape recovery property in water. *J. Mater. Chem.*, 22, 23, 11642–11650, 2012.
3. Pahasup-anan, T., Suwannahong, K., Dechapanya, W., Rangkupan, R., Fabrication and photocatalytic activity of TiO2 composite membranes via simultaneous electrospinning and electrospraying process. *J. Environ. Sci.*, 72, 13–24, 2018.
4. Zhang, R., Wang, X., Song, J., Si, Y., Zhuang, X., Yu, J., Ding, B., *In situ* synthesis of flexible hierarchical TiO2 nanofibrous membranes with enhanced photocatalytic activity. *J. Mater. Chem. A.*, 33, 44, 22136–22144, 2015.
5. Bai, Y., Mao, X., Song, J., Yin, X., Yu, J., Ding, B., Self-standing Ag2O@ YSZ-TiO2 pn nanoheterojunction composite nanofibrous membranes with superior photocatalytic activity. *Comp. Commun.*, 5, 13–18, 2017.
6. Nasir, A.M., Awang, N., Jaafar, J., Ismail, A.F., Othman, M.H.D., A. Rahman, M., Aziz, F., Mat Yajid, M.A., Recent progress on fabrication and application of electrospun nanofibrous photocatalytic membranes for wastewater treatment: A review. *J. Water Process Eng.*, 40, 101878, 2021.
7. Li, Y., Xia, Z., Gong, Q., Liu, X., Yang, Y., Chen, C., Qian, C., Green synthesis of free standing cellulose/graphene oxide/polyaniline aerogel electrode for high-performance flexible all-solid-state supercapacitors. *Nanomaterials*, 10, (8), 2020.
8. Awang, N., Jaafar, J., Ismail, A.F., Othman, M.H., Rahman, M.A., Performance of void-free electrospun SPEEK/cloisite as a function of degree of dispersion state on nanocomposite proton exchange membrane for direct methanol fuel cell application. *Membr.*, 9, (1), 2019.
9. Awang, N., Jaafar, J., Ismail, A.F., Matsuura, T., Othman, M.H.D., Rahman, M.A., Electrospun nanocomposite materials for polymer electrolyte membrane methanol fuel cells BT - organic-inorganic composite polymer electrolyte membranes: Preparation, properties, and fuel cell applications, D. Inamuddin, A. Mohammad, A.M. Asiri (Eds.), pp. 165–191, Springer International Publishing, Cham, 2017.
10. Zheng, Q., Cai, Z., Gong, S., Green synthesis of polyvinyl alcohol (PVA)–cellulose nanofibril (CNF) hybrid aerogels and their use as superabsorbents. *J. Mater. Chem. A*, 2, 9, 3110–3118, 2014.
11. Nogi, M., Kurosaki, F., Yano, H., Takano, M., Preparation of nanofibrillar carbon from chitin nanofibers. *Carbohydr. Polym.*, 81, 4, 919–924, 2010.

12. Wu, Z.-Y., Liang, H.-W., Li, C., Hu, B.-C., Xu, X.-X., Wang, Q., Chen, J.-F., Yu, S.-H., Dyeing bacterial cellulose pellicles for energetic heteroatom doped carbon nanofiber aerogels. *Nano Res.*, 7, 12, 1861–1872, 2014.
13. Li, Q., Zhu, Y.Q., Eichhorn, S.J., Structural supercapacitors using a solid resin electrolyte with carbonized electrospun cellulose/carbon nanotube electrodes. *J. Mater. Sci.*, 53, 20, 14598–14607, 2018.
14. Salazar, H., Martins, P.M., Santos, B., Fernandes, M.M., Reizabal, A., Sebastián, V., Botelho, G., Tavares, C.J., Vilas-Vilela, J.L., Lanceros-Mendez, S., Photocatalytic and antimicrobial multifunctional nanocomposite membranes for emerging pollutants water treatment applications. *Chemosphere*, 250, 126299, 2020.
15. Qayum, A., Wei, J., Li, Q., Chen, D., Jiao, X., Xia, Y., E ffi cient decontamination of multi-component wastewater by hydrophilic electrospun PAN/AgBr/Ag fibrous membrane. *361 (October*, 2018), 1255–1263, 2019.
16. Xie, R., Zhang, L., Liu, H., Xu, H., Zhong, Y., Sui, X., Mao, Z., Construction of CQDs-Bi20TiO32/PAN electrospun fiber membranes and their photocatalytic activity for isoproturon degradation under visible light. *Mater. Res. Bull.*, 94, 7–14, 2017.
17. Kim, B.-H., Kim, C.H., Yang, K.S., Rahy, A., Yang, D.J., Electrospun vanadium pentoxide/carbon nanofiber composites for supercapacitor electrodes. *Electrochim. Acta*, 83, 335–340, 2012.
18. Tai, M.H., Tan, B.Y.L., Juay, J., Sun, D.D., Leckie, J.O., A self-assembled superhydrophobic electrospun carbon–silica nanofiber sponge for selective removal and recovery of oils and organic solvents. *Chem. – A Eur. J.*, 21, 14, 5395–5402, 2015.
19. Panthi, G., Park, S., Chae, S., Kim, T., Chung, H., Hong, S., Park, M., Kim, H., Immobilization of Ag3PO4 nanoparticles on electrospun PAN nanofibers via surface oximation: Bifunctional composite membrane with enhanced photocatalytic and antimicrobial activities. *J. Ind. Eng. Chem.*, 45, 277–286, 2017.
20. Kim, J.H., Mahesh Kumar, J., Joshua, L., Chan Hee, P., Cheol Sang, K., Polydopamine-assisted immobilization of hierarchical zinc oxide nanostructures on electrospun nanofibrous membrane for photocatalysis and antimicrobial activity. *J. Colloid Interface Sci.*, 513, 566–574, 2018.
21. Qayum, A., Wei, J., Li, Q., Chen, D., Jiao, X., Xia, Y., Efficient decontamination of multi-component wastewater by hydrophilic electrospun PAN/AgBr/Ag fibrous membrane. *Chem. Eng. J.*, 361, 1255–1263, 2019.
22. Steckl, A.J., Superhydrophobic and oleophobic fibers by coaxial electrospinning. *Langmuir*, 25, 16, 9454–9462, 2009.
23. Tang, Y., Fu, S., Zhao, K., Teng, L., Xie, G., Fabrication of TiO2 micro-/nanospheres embedded in nanofibers by coaxial electrospinning. *Mater. Res. Bull.*, 78, 11–15, 2016.

24. Virovska, D., Paneva, D., Manolova, N., Rashkov, I., Karashanova, D., Electrospinning/electrospraying vs. electrospinning: A comparative study on the design of poly(l-lactide)/zinc oxide non-woven textile. *Appl. Surf. Sci.*, 311, 842–850, 2014.
25. Ramasundaram, S., Son, A., Seid, M.G., Shim, S., Lee, S.H., Chung, Y.C., Lee, C., Lee, J., Hong, S.W., Photocatalytic applications of paper-like poly(vinylidene fluoride)–titanium dioxide hybrids fabricated using a combination of electrospinning and electrospraying. *J. Hazard. Mater.*, 285, 267–276, 2015.
26. Pant, H.R., Pandeya, D.R., Nam, K.T., Baek, W., Hong, S.T., Kim, H.Y., Photocatalytic and antibacterial properties of a TiO2/nylon-6 electrospun nanocomposite mat containing silver nanoparticles. *J. Hazard. Mater.*, 189, 1, 465–471, 2011.
27. Wang, T., Wei, J., Shi, H., Zhou, M., Zhang, Y., Chen, Q., Zhang, Z., Preparation of electrospun Ag/TiO2 nanotubes with enhanced photocatalytic activity based on water/oil phase separation. *Phys. E Low-Dimens. Syst. Nanostructures*, 86, 103–110, 2017.
28. Ni, Y., Yan, K., Xu, F., Zhong, W., Zhao, Q., Liu, K., Yan, K., Wang, D., Synergistic effect on TiO2 doped poly (vinyl alcohol-co-ethylene) nanofibrous film for filtration and photocatalytic degradation of methylene blue. *Compos. Commun.*, 12, 112–116, 2019.
29. Song, Q., Guo, X., Sun, Y., Yang, M., Anti-solvent precipitation method coupled electrospinning process to produce poorly water-soluble drug-loaded orodispersible films. *AAPS Pharm. Sci. Tech.*, 20, 7, 273, 2019.
30. Robinson, T.M., Hutmacher, D.W., Dalton, P.D., The next frontier in melt electrospinning: Taming the jet. *Adv. Funct. Mater.*, 29, 44, 1904664, 2019.

Index

Also of Interest

Check out these other books by Shahid-ul-Islam published by Scrivener Publishing

Green Chemistry for Sustainable Water Purification
Edited by Shahid-ul-Islam, Abid Hussain Shalla and Mohammad Shahadat
Published 2023. ISBN 978-1-119-85229-2

Role of Microbes in Industrial Products and Processes
Edited by Sanjay Kumar, Narendra Kumar Shahid-ul-Islam
Published 2022. ISBN 978-1-119-90119-8

Handbook of Biomass Valorization for Industrial Applications
Edited Shahid-ul-Islam, Aabid Hussain Shalla and Salman Ahmad Khan
Published 2021. ISBN 978-1-119-81873-1

Chemistry of Biologically Potent Natural Products and Synthetic Compounds
Edited by Shahid-ul-Islam and Javid Ahmad Banday
Published 2021. ISBN 978-1-119-64034-9

Advanced Functional Textiles and Polymers
Fabrication, Processing and Applications
Edited by Shahid-ul-Islam and B.S. Butola
Published 2020. ISBN 978-1-119-60579-9

Advances in Metallodrugs
Preparation and Applications in Medicinal Chemistry
Edited by Shahid-ul-Islam, Athar Adil Hashmi and Salman Ahmad Khan
Published 2020. ISBN 978-1-119-64042-4

Environmental Nanotechnology for Water Purification
Edited by Shahid-ul-Islam
Published 2020. ISBN 978-1-119-64045-5

Integrating Green Chemistry and Sustainable Engineering
Edited by Shahid-ul-Islam
Published 2019. ISBN 978-1-119-50982-0

Advanced Textile Engineering Materials
Edited by Shahid-ul-Islam and B.S. Butola
Published 2018. ISBN 978-1-119-48785-2

Nanomaterials in the Wet Processing of Textiles
Edited by Shahid-ul-Islam and B.S. Butola
Published 2018. ISBN 978-1-119-45980-4

Advanced Materials for Wastewater Treatment
Edited by Shahid-ul-Islam
Published 2017. ISBN 978-1-119-40776-8

Plant-Based Natural Products
Derivatives and Applications
Edited by Shahid-ul-Islam
Published 2017. ISBN 978-1-119-42383-6

Printed and bound by CPI Group (UK) Ltd, Croydon, CR0 4YY
17/09/2026
14964274-0004